NIMBUS Maintenance

English edition - 2018

Knud Jørgensen
Ben Geutskens and Charles Duffill

NIMBUS
Maintenance

English edition - 2018

First published in 1990 by Notabene, Copenhagen DK
Danish edition 2008: *Nimbus – vedligeholdelse* by Motorploven, Hadsten DK

A shortened English edition 2012 was published in 2012 by Books on Demand GmbH, København, Danmark
This new size edition is published in 2018
Cover photo: Finn Nielsen

Publisher: BoD – Hellerup, Denmark

Printing: BoD – Norderstedt, Germany

ISBN: 9 788 743 003 724

© Knud Jørgensen

All rights reserved. With exception of quoting brief passages for the purpose of review, no part of this publication may be recorded, reproduced or transmitted by any means, including photocopying without permission of the author (exceptions are given in the book).

FOREWORD to the English edition 2018

The first edition of *'NIMBUS – vedligeholdelse'* (NIMBUS - Maintenance) was published in spring 1990 and at that time titled: *'NIMBUS - og kunsten at vedligeholde den'*. (NIMBUS: The Art of Maintenance).

A second Danish edition was published in 2008, and many have asked for an English edition, hence the publication of this edition in 2012. This new size edition was published 2018

There were 4,500 Nimbus-C motorcycles registered in Denmark in spring 2015. If we take account of the 1,000 in use abroad, almost half of the total production of 12,715 motorcycles survive to this day. All require maintenance and repair, and from time to time renovation and restoration. This book is about the first three of these subjects, and together with *'Nimbus -Technical Development'*, it may also be useful in restoration. The content of this book is by large limited to those operations which a skilled owner can do or can have done. Many repairs nowadays have to be left to a professional workshop, as mistakes can become very expensive and irreplaceable original parts may be damaged.

Many thanks to Ben Geutskens and to Charles Duffill for their work on the English edition. Errors or shortcomings remain the author's responsibility.

Højbjerg, October 2018

Knud Jørgensen

Introduction

This book is intended as a guide to the maintenance and repair of Nimbus type C motorcycles built between 1935 and 1959. The first Nimbus-C series, built during 1934 and 1935 and with serial numbers 1301-1551, differ somewhat from later models. See *Andersen, J. B. (1996): 'Nimbus model C 1934'*. An English translation of this book is available on www.geutskens.eu
The 1934 rear wheel to which this book refers was in fact fitted up to 1937, and the 1934-2 carburettor model was used up to 1938.

The earlier Nimbus type A/B 1918-1928, widely but unofficially known as the 'Kakkelovnsrør' or 'Stovepipe', is described in *Andersen og Jørgensen (2007). 'Nimbus 1918-28 "Kakkelovnsrøret"*

The many changes to Nimbus-C between 1934 and 1959 are described in *Jørgensen, K. (2016): 'Nimbus – Technical Development 1934 - 1959'*, This book explains the 'early' versus the 'later' design, the 'low' telescopic fork as opposed to the 'high' version, etc.

In the description of the various dismantling, repair and reassembly procedures, spare part numbers are given in brackets. The spare parts catalogue issued by A/S Fisker & Nielsen in 1951 and corrected in 1958: (*'Værkstedshåndbog – originale reservedele'*) has been reprinted many times by Denmark's Nimbus Touring Club (DNT) and others.

Every illustrated part in the catalogue is numbered. This part number will be found in the list on the opposite page. An underscored illustration number indicates that the numbers of any earlier versions of the part are also listed. The serial numbers applicable to different versions are given in the column to the left of the part numbers.

Reference is also made to *Weidinger, S. (2007): Nimbus – Ordbog – Dictionary - Wörterbuch.*

Contents

Foreword 5
    - Introduction 6

Contents 8
Specifications 11

Workshop practice ... 13
    - torque settings........ 211

Engine
    - Removal 23
    - Installation............ 27
    - Installed engine work.. 31
    - Dismantling 36
    - Assembling major
      components 41
    - Final assembly 53

Engine checklists
    - Cylinder block 72
    - Cylinder head 74
    - Crankshaft 78
    - Pistons 80
    - Camshaft 82
    - Connecting rods 84
    - Clutch 86

Electrics ....
    - Battery 89
    - Dynamo 91
      maintenance
      repair and testing 94
    - Voltage regulator .. 99
    - Ignition coil and
      distributor 103

    - Combination switch .. 105
   -Ammeter and
      charge warning light .. 106
    - Horn and
      horn button 107
    - Brake light switch ... 109
    - Lighting equipment
        - Headlight 110
        - Tail light 111
        - Sidecar lights ... 113
    - Wiring 113
    - Fuses 117
    - Instrument lighting 118

Gearbox
    - Maintenance 119
    - Repair 119

Drive shaft 121

Carburettor
    - Maintenance 124
    - Dismantling and cleaning 123
    - Adjustment
      carburettor 1934-1... 126
      carburettor 1934-2 . 128
      carburettor 1938 128
      carburettor 1950 129
      carburettor 1951 131
      carburettor 1951-2 . 131
      carburettor 1953 ... 132

Wheels
    - Hubs 134
    - Spokes 134

- Nipples	135
- Rims	135
- Checking wheel truth	135
- Wheel-building and 'trueing'	136
- Emergency repair	137
- Tyre pressures	210
- Front wheel	
- Fitting front wheel	139
- Dismantling	140
- Assembling	141
- Speedometer gearbox	142
- Rear wheel	
- Removing	143
- Dismantling	144
- Initial assembly and adjustment	146
- Final drive pinion	148
- Crown wheel and pinion adjustment	152

Brakes
- 150mm brakes	155
- 180mm brakes	157
- Maintenance and repair.	158
- Adjusting the brakes	
- front brake	160
- rear brake	161
- sidecar brake	161

Frame
- Frame straightening	162
- Frame modifications	163
- Steering lock	165
- Knee pads	165
- Footrests	165
- Tool box	166
- Centre stand	166

Petrol tank
- Removing	167
- Fitting	167
- Repairing	168

Telescopic forks
- Removing the complete fork assembly	169
- Fitting the forks	170
- Preparation, assembly and fitting	170
- Dismantling the forks	171
- Assembling the forks	175
- Fork repair	182

Handlebar 183

Saddle and pillion seat
-Coil-sprung saddle	189
- Rubber suspended saddle	191
-Coil-sprung pillion seat	191
-Rubber-suspended pillion seat	192

Gear change mechanism
- Removing and fitting	193
- Dismantling and assembling	194
- Repairing	195
- Adjusting	196

Speedometer
- Removing and fitting	197
- Dismantling	199
- Maintenance and repair	199

Valve enclosure 200

Sidecars
- Sidecar types..................... 200
- Maintenance and repair.... 201
- Adjusting the brake.......... 202

Adjustments
- Rear wheel bearings......... 203
- Brakes............................. 143
- Crown wheel and pinion (final drive gears) 148
- Carburettor...................... 126
- Valve clearances............. 203
- Contact breaker............... 206
- Ignition timing................ 207
- Valve timing 208
- Camshaft and dynamo bevel gears................. 207
- Helical cut bevel gears.... 207
- Tyre pressures................. 210
- Torque settings................ 210
- Lubrication...................... 211
- Paint Colours 212

Specifications for Nimbus-C

Number of cylinders: 4 in line
Bore: 60mm
(available oversize's: 60.6mm, 61.2mm or 61.8mm)
Stroke: 66mm
Cubic capacity: 746 cm³ (761 cm³, 776 cm³ or 792 cm³).
Figures in brackets are for 1st, 2nd and 3rd oversize.
Compression ratio: 5.7:1 (domed pistons) or 5.4:1 (flat-top pistons)
Power output: 22 bhp / 16.2 kw @ 4500 rpm. (5.7:1 compression)
18 bhp / 13.2 kw @ 4500 rpm. (5.4:1 compression)
Max. speed: Solo: 120 km/h.; with sidecar: 95 km/h.
Cruising speed: Solo: 90 km/h.; with sidecar: 75 km/h.
Gear ratios: 1st: 2.43:1, 2nd: 1.53:1 (or 1.57:1), 3rd: 1.00:1 (or 1.04:1)
Final drive ratio: Sidecar gearing: 4.92:1 (59/12)
Solo vehicle: 4.07: 1 (57/14) or 4.00 (56/14)

Brakes: 150mm or 180 mm drums, front and rear
Wheel base: 1301 – 7500: 1410mm
7501 – 14015: 1435mm

Rake: 65 °

	Frames 1301 – 7500	*Frames 7501 – 14015*
Trail:	60mm	65mm
Overall lenght:	2160mm	2200mm
Overall width:	780mm	720mm
Overall height:	1050mm	1100mm

Seat height: 710mm

Dry weight:	185kg (with passenger seat)
Petrol tank:	12,5 litre, of which approximately 1 litre reserve
Tyres:	3.50" x 19" (possible front wheel »Sport« 3,25" x 19")
Wheel base:	1301-7500: 1410mm; 7501-14 015: 1435mm
Front fork offset:	1301-7500: 60mm; 7501-14 015: 65mm
Valve clearance:	Inlet: 0.3 mm, exhaust: 0.7 mm (cold engine)
Contact breaker gap setting:	0.7mm
Plug gap setting:	0.7mm
Ignition sequence:	1 - 3 - 4 - 2
Ignition timing:	1650 rpm: 37° before top dead centre (TDC)

Valve timing:	*Inlet valve*
	    opens: 7° before TDC
	    closes: 39° after bottom dead centre (BDC)
	 Exhaust valve
	    opens: 42° before BDC
	    closes: 4° after TDC

Battery:	6 V/ 12 Ah
Dynamo:	8V/70W
Lighting:	Low beam / high beam: 35/35 W Parking light: 4 W (possible 5 W) Tail light and licence plate light: 5 W Brake light: 15W Instrument light: 1,2W Ignition/charge warning light (Bosch): 4W Ignition/charge warning light (lighting switch): 2W

Workshop practice

- Give a task all the time it needs.
- Arrange good lighting.
- Have pencil and paper within reach.
- Have plenty of clean rags available.
- Use service tools where necessary and quality tools in all other cases.

Fasteners and locking methods:

Thread types

All threads on the Nimbus-C are right hand, with the exception of that on the speedometer drive worm (7572 or 8361), which is a left-hand thread. With few exceptions, threads are standard metric or metric fine.

Fasteners

Bolts, screws, washers and nuts for Nimbus-C are standard engineering hardware items. Their precise sizes can generally be read from the spare parts list. For example: Plan 9C in the spare parts list calls for: *7651 Screw 6 – 0,75 x 13mm,* that is, an M6 screw with an under-head length of 13mm and a thread pitch of 0.75mm (metric fine).

Bolts

On the Nimbus-C all bolts are hexagon head, and the following sizes are used:
* 6mm (10mm across flats)
* 8mm (originally 14mm across flats, replacements 13mm)
* 10mm (17mm across flats)
* 12mm (19mm across flats)

Screws

All screws are slot head type.

Studs

All studs are 6mm, 8mm or 12mm, either plain or shouldered

Nuts

All nuts are hexagonal, either plain, castellated (locked by split pin) or domed (as on wheel spindles). (Dynamo bevel gear retaining nuts are an exception, being round castellated pattern requiring a special spanner).

 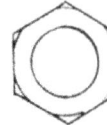 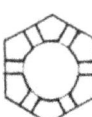 

Washers

Fibre washers are used in some places to prevent damage to surfaces. Flat washers of galvanised or stainless steel are used for the same purpose. *Spring washers* are locking washers (see below).

Locking methods

* *Spring washers* are used when connecting parts. Spring washers must be in direct contact with the screw head or nut. A flat washer should be placed under a spring washer if the fastening to a light-alloy metal surface,
* *Locking washers / locking tabs* prevent fasteners from loosening, typically by means of a section of the washer or tab being folded against a flat of a bolt head or nut, while the washer or tab is itself secured against turning.

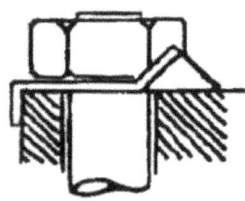

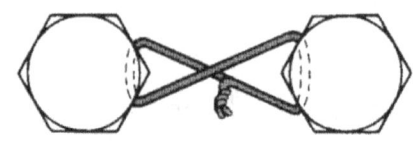

* *Locking wire* can be used if the bolt head or nut is drilled.

* *Split pins* are used to lock castellated nuts to drilled bolts.

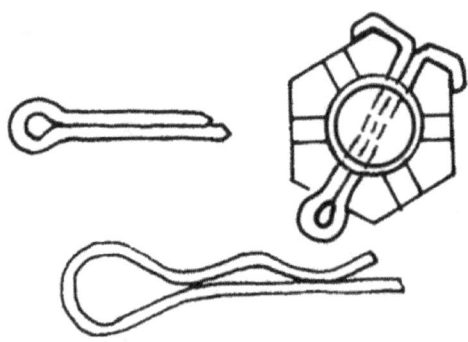

* *Special purpose adhesives* such as 'Loctite' are now available and are very effective when applied to thoroughly cleaned and dry fasteners. There are various grades of these 'anaerobic expansive' products. Some are suitable for locking bearings into place.

Torque settings

In some cases, applying the correct torque setting to a fastener is essential. (Page 210) The head of the bolt may bear a code indicating the grade of material and therefore the maximum allowable torque. Recommended torque settings assume clean, dry, undamaged threads. Torque figures are usually specified in foot-pounds (ft/lbs) or Newton-metres (Nm).

Hand Tools

Hammers

* Engineer's ball-peen hammer
* Soft-faced hammer (rubber, plastic, or fibre faced)

Spanners

* Adjustable spanners
* Ring spanners
* Open-end spanners
* Combination spanners (one ring end and one open end of the same size)

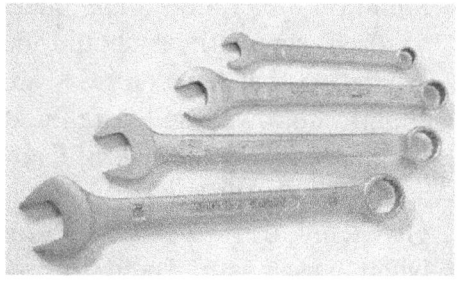

* Sockets - used with square-drive ratchet handles
* Box spanner / tubular spanner
* Torque wrench

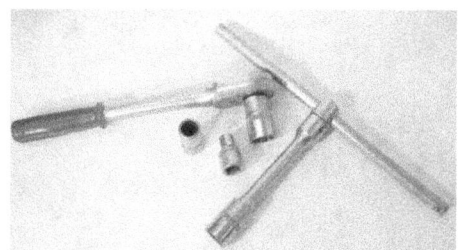

Pliers

* Combination pliers
* Circlip pliers
* Multigrip pliers

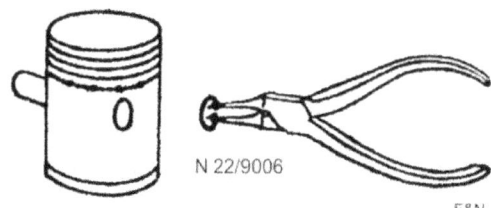

Screwdrivers

Make sure you have a selection of flat-blade screwdrivers to fit the varied sizes of slot-head screws found on the Nimbus. The blade of each screwdriver should be ground so that the flat sides are slightly hollowed and parallel at the tip, reducing the risk of damaging the screw head. Right-angled screw drivers can be useful where access is difficult (for example when removing the Nimbus dynamo's brush holder screws).

Special-purpose tools

For the Nimbus-C motorcycle the factory developed a set of 'Service Tools' for certain specific repair operations. General-purpose tools can be used, but with care and less conveniently for some, but not all of these operations. In such cases, repairs have to be carried out in a specialised workshop. Factory service tools marked * appear in the DNT Drawing Archive.

Where service tools are mentioned or illustrated in this book, tool numbers and the Archive drawing numbers are given.

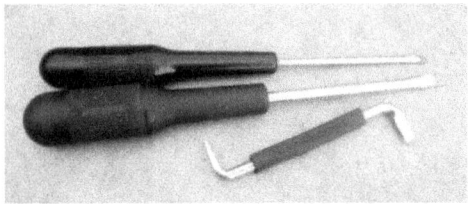

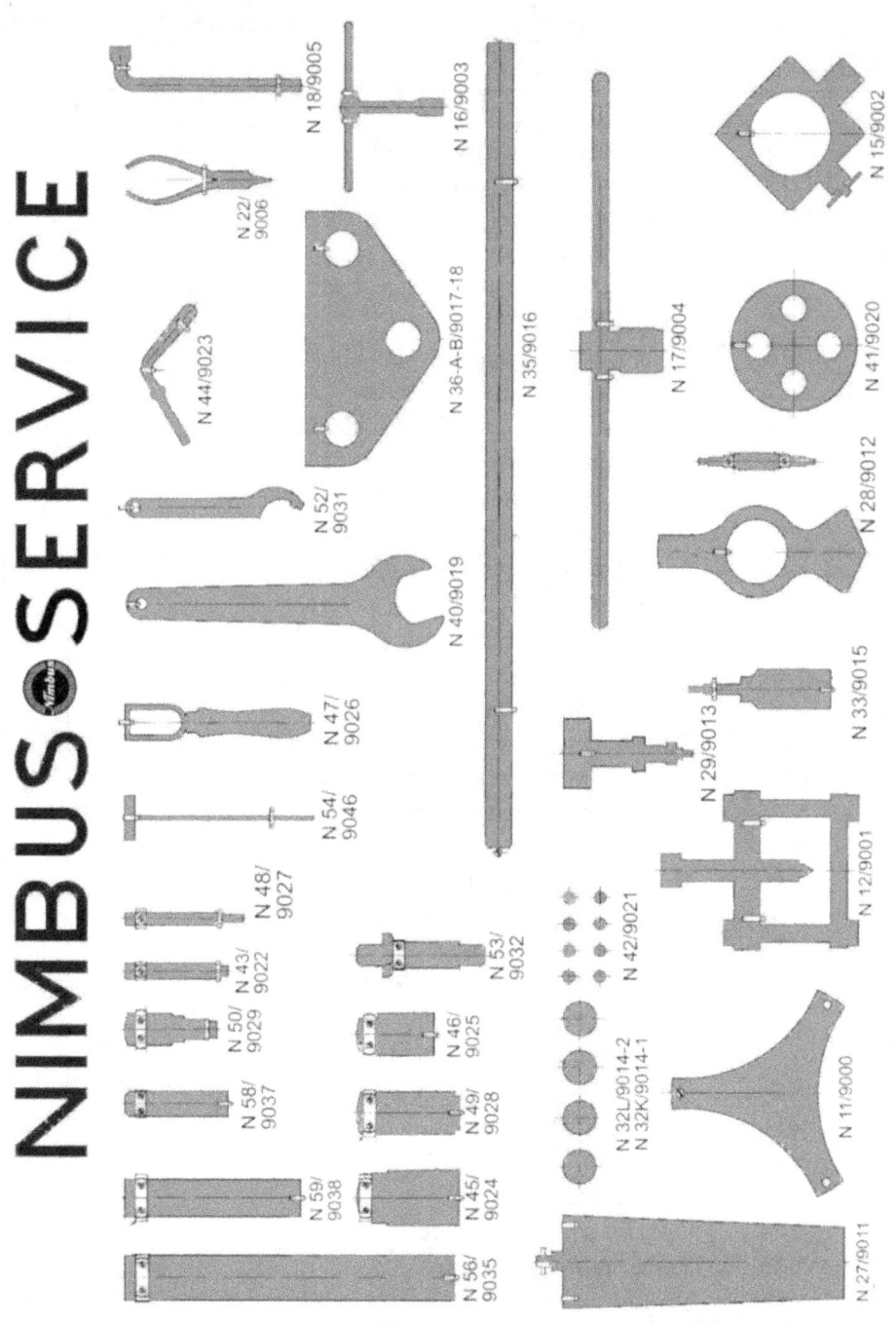

Drawing	Tool	Tool description
8999	N10	Puller for dynamo gear wheel
8999	N14	Puller for dynamo gear wheel
9000*	N11	Puller for flywheel
9000-2*		Puller for flywheel
9001*	N12	Puller for crankshaft bevel gear wheel and main bearing
9002*	N15	Holding bracket for dynamo
9003*	N16	T-bar socket for upper dynamo bevel gear retaining nut
9004*	N17	T-bar socket 27mm for flywheel and fork stem top nut
9005*	N18	Offset 14mm spanner for cylinder head fasteners
9006	N22	Circlip fitting tool
9007		Reamer, connecting rod small end bush
9007		Puller for camshaft bevel gear wheel
9008	N25	Valve facing set, complete in box
9009	N26	Engine stand
9010		Workshop stand
9011*	N27	Drive shaft tool
9011-2		Puller for drive shaft hub
9012*	N28	Adjustment gauge for final drive pinion
9013*	N29	Puller for wheel bearing outer race
9014*	N32	Distance piece, rear hub assembly
9015*	N33	Fork bush extractor for bush
9016*	N35	Test mandrel for fork tubes
9017*	N36A	Test gauge for fork tubes
9018*	N36B	Test gauge for fork tubes
9019*	N40	Spanner 46mm steering head bearing nut
9020*	N41	Alignment tool for clutch plate
9020-2*		Alignment tool for clutch plate
9021*	N42	Retainer for connecting rod bolt
9022*	N43	Drift for gudgeon pin
9023*	N44	Clutch compression tool
9024*	N45	Drift for main bearing
9025*	N46	Drift for bevel gear wheel
9026*	N47	Valve spring compressor
9027*	N48	Drift for valve guide

9028*	N49	Drift for small end bush and bevel gear bush
9029*	N50	Drift for small end bush
9030*		Frame alignment gauge
9031*	N52	'C' spanner for speedometer worm gear
9032*	N53	Drift for front wheel bearing and telescopic fork bushes
G-9033*	N29	Extractor for front wheel bearing
9034*		Rear frame alignment tool
9035*	N56	Drift for fork stem bearing cone
9036		Drilling template for cylinder head (Valve enclosure)
9037*	N58	Drift for removing sidecar axle
9038*	N59	Drift for fitting sidecar axle
9040*		Tool board with profiles
9041*		Tool holders for tool board
9042*		Tool board
9043*		Installation tool for dynamo upper bevel gear 7115
9044*		Installation tool for dynamo upper bevel gear 7147
9046*	N54	Hand reamer 16mm for connecting rod small end bush
9047*		Drilling template for saddle suspension brackets
9048		T- bar socket 14mm
9049		T-bar socket 19mm

Improvised tools

If factory service tools are not available, consult the Drawing Archive. In many cases it may be possible to make a substitute, as exact dimensions are given.
Right: In absence of tool number 9003/N16 T-bar socket.

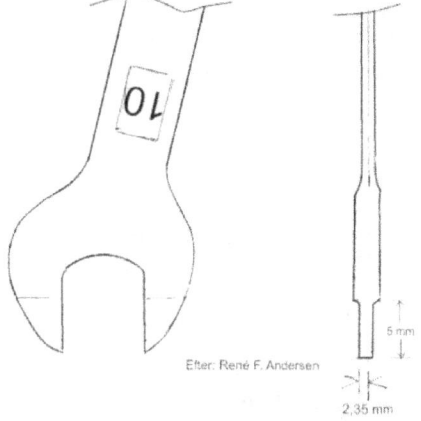

Efter: René F. Andersen

Measuring equipment

In order to be able to take dimensions in case of engine overhaul, a vernier, dial, or digital calliper and a micrometer will be needed. Dial gauges are particularly useful for measuring specified tolerances (as in trueing wheels) or backlash (when fitting gears).

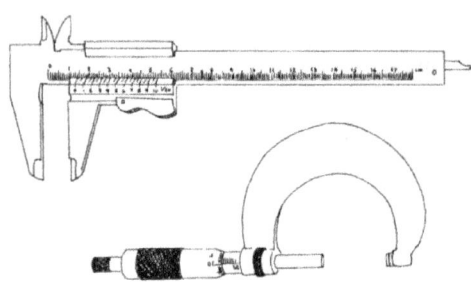

Drawing Archive

The Nimbus-C factory engineering drawings available in the Drawing Archive were produced 40 to 70 years ago to the standards of that time, and are certainly very useful today. Some more recent drawings lack precise spare part dimensions and do not comply with the requirements for technical drawings. Some spare parts do not have original factory part numbers but have been assigned an identifying number. That is the case with helical timing gears for example.

In 2006 the Danish Nimbus Touring Club (DNT) produced two compact discs with original Nimbus construction drawings and drawings of prototypes and new developments.

The opening text on the CD reads:

Drawings from A/S Fisker and Nielsen's drawing archive were registered and photocopied by Danmarks Nimbus Touring in 1982 and as many missing drawings as possible have been re-created by measuring spare parts. All drawings were scanned by

Danmarks Nimbus Touring between 2002 and 2005.

*The Nimbus-C drawings are organized according to A/S Fisker and Nielsen's Parts Catalogue published in 1951 and updated in February 1958. Drawings for the Nimbus A/B are organised according to component categories. Drawing numbers with a preceding **G** are combined drawings which illustrate multiple parts. The meaning of a preceding **E** is unclear, but might indicate a repeat order. Trailing characters to a drawing number indicate the following:*

__S__ (Smede) - the drawing is of forged parts
__P__ (Plade) - the drawing is of sheet metal parts
__A__ (Arbejde) - a working or production drawing
__K__ (Kalkule) - a drawing including calculations or other documentation.

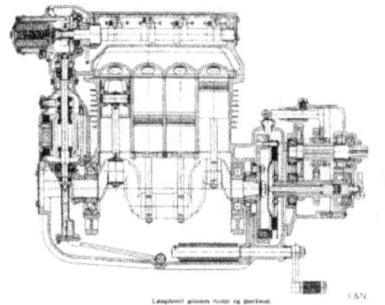

Nimbus-C archive drawings

Whereas there is a limited collection of drawings for the Nimbus-A/B, an almost complete set of drawings for the Nimbus-C survives. The drawings are on CD and arranged in folders corresponding to the organisation of the Nimbus-C Parts Catalogue. Each folder contains all drawings

Each file contains a sub file with drawing in PDF-format (to be opened with Adobe Reader) and a file with drawings in TIF-format. Besides, all files open up automatically in JPG-format. Most of the scanned drawings are very large and besides that, very dirty. That means that a certain computer capacity is required for opening and printing the drawings, where required, especially those which are in TIF-format.

Archive searching

Use the computer search function, specifying the drawing number (part number).

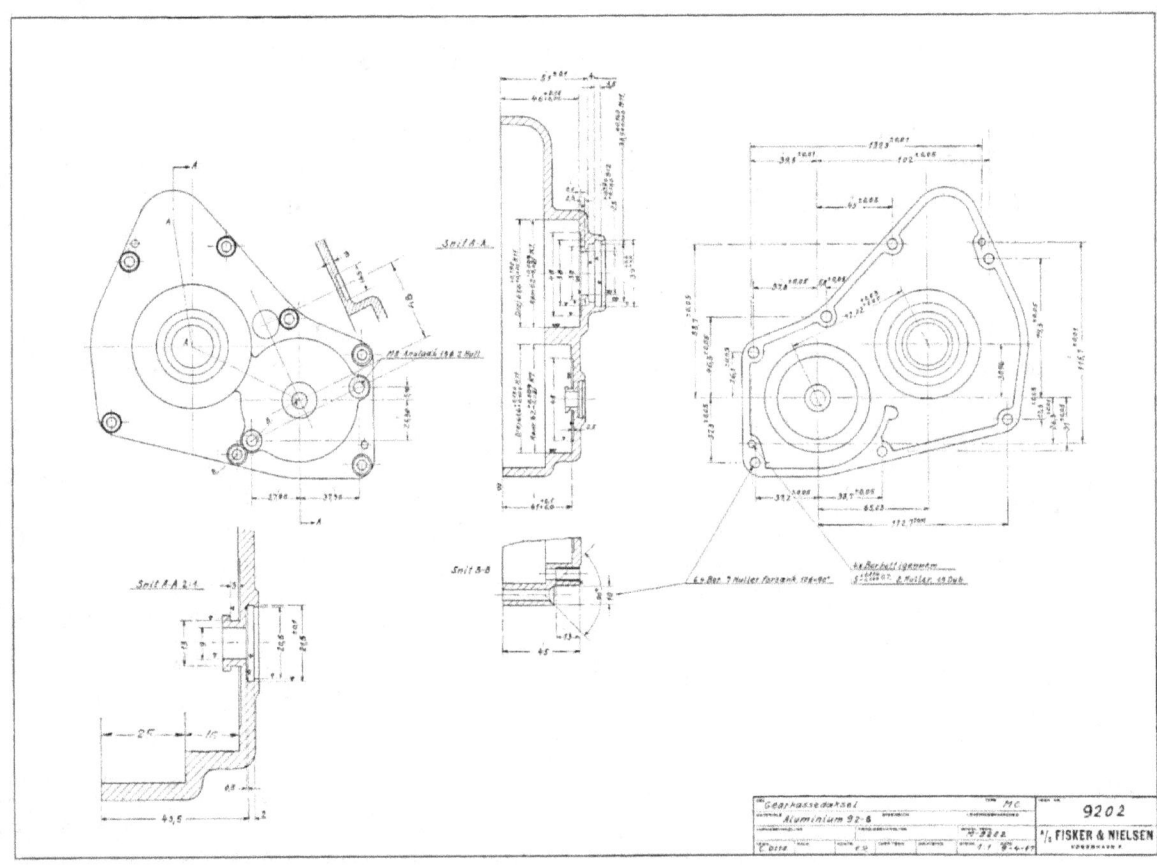

Engine

Removing the engine from the frame

Disconnect all wires from the battery. Disconnect the earth/fuse holder (8032) from the engine (where applicable).

Remove the screws (5400) from the brush-holder cover (7872) on the right side of the dynamo. Label the 'D' and 'F' wires before disconnecting them to avoid any mistake when reconnecting them. Disconnect the wire to the 'D' carbon brush and remove the brush. Disconnect the 'F' wire.

Disconnect the 'T' wire from the ignition coil (7706) (see wiring diagram).

Remove the ignition coil by detaching the H.T. leads, swinging down the retaining clip (7690 or 8185), and easing the ignition coil forward. If the coil does not readily separate from the distributor (7392 or 8175), remove the screw (5350) from the distributor adjustment plate and remove the ignition coil and distributor together. This will remove the rotor (7432) at the same time.

- Remove the carburettor:

- Remove the crankcase breather pipe (7684 or 8583), which also serves as the oil filler cap. Block the oil filler hole with a clean rag or similar.

- Detach the throttle cable (7673 or 8927) from the hook of the twist grip and free the cable from the frame.

- Remove the fuel line (7683, 8183 or 8577).

- Remove the air filter (8584), which is attached by three screws (8578). *Note:* The 1934 carburettor has an air intake screen (7498) which is removed together with the carburettor.

- Remove the choke assembly (8597) and the carburettor. These are secured to the cylinder head inlet manifold by the same two screws (7429 or 8594).

- Remove the paper gasket (7521).

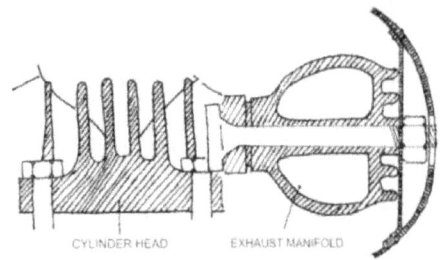

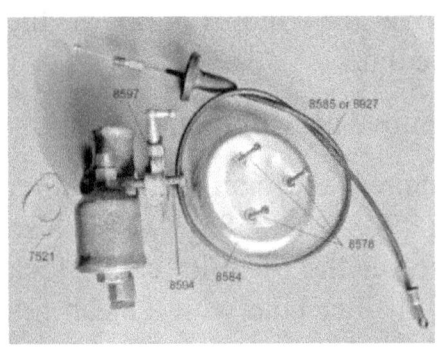

- Remove the exhaust manifold (7144) and heat shield (8023 or 8454). *Take care:* The bolts and nuts will often be tight due to corrosion. Note that different fixing arrangements for the manifold and the heat shield have been used. The exhaust pipe can also be removed at this stage.

- Disconnect the clutch release and gearchange mechanisms.

For hand-change machines:

- Remove the pivot bolt (7270) and remove the gear lever (7477) from the frame. Remove the pull rod (8408) that connects the clutch pedal to gearbox.

- Remove the split pin (3867) securing the clutch cable to the clutch lever (8415) and release the cable (8413).

- Remove the lever pivot bolt (8416) and the lever.

- Remove the linkage (8408) which connects the foot change assembly to the gearbox.

For foot-change and later version gearbox machines:

- Pull in the clutch until a wooden block can be wedged between the clutch release arm (9218) and the mudguard.

Remove the split pin (3867) from the release arm and remove the clutch cable (8944). Wire the release arm in place and then remove the block.

- Remove the linkage (9221) from the foot-change assembly to the gearbox.

- Remove the tool box (7610) if it is mounted under the frame.
Remove the centre stand (7300 or 8328). The motorcycle will need to be supported by blocks under the footrests.

- Pack blocks under the sump to take all weight off the wheels.

- Remove the four bolts (7269) which secure the engine to the frame.

There are alternative ways to continue:

Either:

- Remove the rear wheel.

- Remove the drive shaft (7131, 8258 or 8258-2) and its compression spring (8388) from the lay shaft.

- Replace the rear wheel, but without the drive shaft. Do not tighten the bolts that secure the brake back plate and final drive housing to the rear wheel and do not tighten the nuts of the rear axle.

- Lift the frame up and forward over the engine. Note that the centre stand-mounting brackets of the lower frame rail have to be eased apart slightly to allow the centre stand spring attachments to clear the crankcase - sump joint.

- The centre stand can now be refitted to the frame.

Or:

Leave the rear wheel in place.

- Ease the motorcycle back until the drive shaft (7131, 8258 or 8258-2) can be disengaged from the lay shaft while remaining fully engaged at the final drive assembly. Recover the compression spring (8388).

- Lift the motorcycle by the fork legs and move it back over the engine. Note that the centre stand-mounting brackets at the lower end of the frame have to be eased apart slightly to allow the two pins for the centre stand springs to clear the crankcase - sump joint.

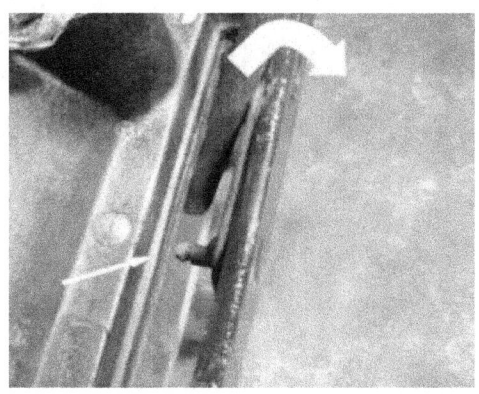

- The centre stand can now be refitted to the frame.

Replacing the engine in the frame

- Place the engine and gearbox assembly on blocks, so that the four engine mounting holes will be at the same height as the mounting holes in the lower frame when both wheels are fitted to the motorcycle.

For an engine with the later gearbox, the clutch release arm (9218) has to be wired in the fully-in position so as to clear the mudguard.

- Lift the frame over the engine and lower it into place. Without the rear wheel, and with the rear mudguard hinged up, move the frame forward and down. Alternatively, with the rear wheel in place, move the frame back and down. In both cases the centre stand mounting brackets of the lower frame rail have to be eased apart slightly to allow the centre stand spring attachments to clear the crankcase-sump joint.

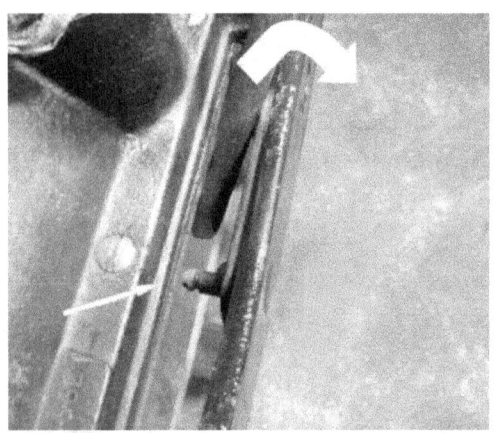

- Fit the compression spring (8388) to the layshaft before sliding the drive shaft (7131, 8558 or 8558-2) into place.

- Fit the rear wheel if it was removed. Relocate the rear mudguard supports onto the rear wheel spindle.

- Move the frame forward so that the drive shaft is fully engaged with the splines of the final drive pinion wheel. Align the engine mounting holes in the frame and sump.

- Fit the four engine mounting bolts (7269) and lock washers. Note that the bolts have 26mm heads. Tighten the four bolts by stages, in turn.

- Support the motorcycle by blocks under the footrests and remove the blocks from under the sump.

- Fit the centre stand (7300 or 8328) between the stand attachment brackets and with the cross-brace on the underside. Fit the pivot bolts (7273 or 8329) from the inside so that the lock washers and nuts (7260) are outside the attachment brackets. Do not tighten the nuts until the centre stand springs (7585) have been fitted. This can be done in several ways. One way is to locate each spring on the stand and then to pull the other end of the spring up to the frame attachment point using a cord or wire passed up between the engine and the frame.

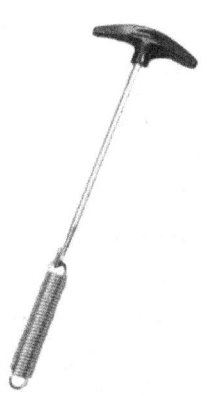

- Put the motorcycle on its centre stand and tighten the stand pivot nuts, mudguard hinge nut, all rear wheel attachment bolts and rear spindle nuts.

- Connect the clutch release and gearchange mechanisms:

For hand-change machines:

- Fit the clevis pin (7254) to the clutch pedal (7281) and pass the threaded end of the pull rod (7670) through it.

- Fit the pull rod to the clutch release arm (7076) and fit the split pin (3885)

- Fit and adjust the adjusting nut (7256) of the pull rod.

- Pass the gear lever (7477) through the slot of the frame, position the lower end in the slot of the shaft and use the pivot bolt (7270) to secure the lever to the pivot bolt bracket.

- Run the free end of the clutch cable (7672) through the guide in the lower part of the frame and fit it to the clutch pedal using the clevis pin (7479) and split pin (3866).

- Adjust the cable by means of the cable adjuster (7208) at the clutch lever.

For foot-change and <u>early</u> gearbox machines:

- Pass the lever (8415) through the hole of the clutch release arm and fit the pivot bolt (8416) through the threaded hole in the lower frame rail so that it engages with the hole in the lever. Tighten the pivot bolt.

- Pass the free end of the clutch cable (8413) through the cable abutment on the foot-change assembly and fit the eye of the cable over the lever. Fit the split pin (3867). Free the clutch release arm.

- Fit the linkage (8408) between the foot-change assembly and the gear selector shaft.

For foot-change and <u>later</u> gearbox machines:

- Pass the free end of the clutch cable (8944) through the cable abutment on the gearbox and fit the eye of the cable over the lever (9218). Fit the split pin (3867). Free the clutch release arm.

- Fit the linkage (9221) between the foot-change assembly and the gear selector shaft.

- Fit the exhaust manifold and heat shield.

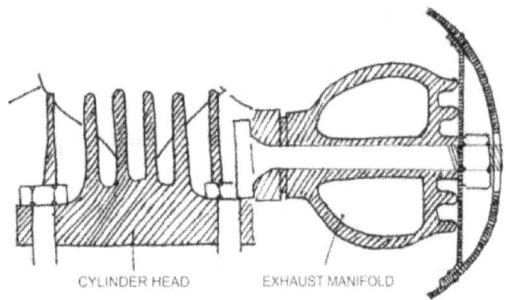

- Fit the carburettor.

- Fit the ignition coil (8054 or 7706). If the distributor has also been removed, fit this first. Ensure that the contact prong of the coil enters the corresponding hole in the distributor and slide the coil home, taking care that the rotor button (8055) is not damaged or displaced as it contacts the rotor (7432). The contact breaker gap setting and ignition timing have to be adjusted.

- Fit the ignition coil retaining clip (7690 or 8185).

- Fit the 'I' lead to the ignition coil (green cable - see wiring diagram).

- Fit 'D' and 'F' leads (blue and yellow cables) to the D and F terminals of the right-side dynamo brush holder. The carbon brush lead must also be connected to Terminal 'D'.

Take care: If there is any doubt as to how the cables have to be fitted, check all connections with a test lamp first. Fitting the cables incorrectly will damage the voltage regulator.

- Fit the tool box (7610) so that the lid is to the right side of the motorcycle.

Operations possible with the engine in the frame

With the engine still fitted in the frame it is possible to remove the following components:
* Carburettor
* Exhaust manifold and heat shield
* Camshaft housing
* Dynamo
* Cylinder head
* Gearbox - after removal of the engine-gearbox bolts (7269).

If the engine has to be totally dismantled, or if the oil pump, crankshaft, connecting rods, pistons, kick-starter or clutch have to be repaired, the engine has to be removed from the frame.

Removing the carburettor

Remove the crankcase breather pipe (7684 or 8583), which also serves as the oil filler cap. Block the oil filler hole with a clean rag or similar.

- Detach the throttle cable (7673 or 8927) from the hook of the twist grip and free the cable from the frame.

- Remove the fuel line (7683, 8183 or 8577).

- Remove the air filter (8584), which is attached by three screws (8578). *Note:* The 1934 carburettor has an air intake screen (7498) which is removed together with the carburettor.

- Remove the choke assembly (8597) and the carburettor. These are secured to the cylinder head inlet manifold by the same two screws (7429 or 8594).

- Remove the paper gasket (7521).

Removing the exhaust manifold and heat shield

- Remove the fasteners attaching the exhaust manifold (7144) and heat shield (8023 or 8454) to the cylinder head. *Note:* Different exhaust manifold and heat shield fittings have been used. Because of large variations in temperature, fasteners may have corroded. Use penetrating oil and work carefully, if parts are to be re-used.

- Lift the exhaust manifold and heat shield away from the cylinder head and exhaust pipe.

- Remove the 4 exhaust port gaskets (7645).

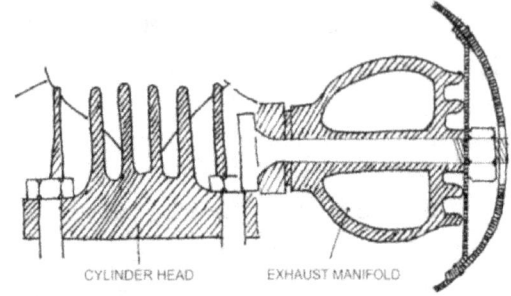

Removing the camshaft housing

- Disconnect the 'I' lead from the ignition coil. Identify it to ensure correct replacement.

- Remove the ignition coil retaining clip (7690 or 8185), ignition coil (8054 or 7706), rotor (7432) and distributor (7392 or 8175). (See *Removing the engine from the frame*)

- Remove the camshaft oil feed and return pipe (7865) and the paper gasket (7517).

- Remove the 4 nuts (7258), which hold the seal clamping ring (7463) to the camshaft housing. -

- Release the ring by pushing it down. *Take care:* Avoid damaging the threads of the studs.

- Slacken the 4 bolts (7190) or nuts (7259) which secure the camshaft housing to the cylinder block, gradually and in turn.

Take care: Remove the 2 rear fasteners first, to avoid damaging the rear camshaft housing mounts.

- Remove the spark plug caps (7510).

- Remove the fuel tank.

- Lift the camshaft housing clear of the dynamo and cylinder head.

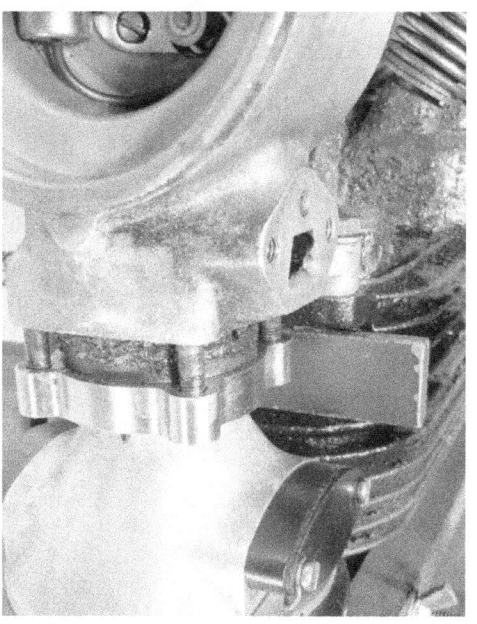

Removing the dynamo

- Remove the camshaft housing.

- Remove the brush holder cover screws (5400) and the cover on the right side of the dynamo. Label the 'D' and 'F' leads before disconnecting to ensure correct reconnection.

- Disconnect the brush lead at the 'D' terminal and remove the brush. Disconnect the 'F' lead.

- Remove the 4 bolts (7191) securing the dynamo to the cylinder block.

- Lift the dynamo from the cylinder block. If any sealant makes this difficult, tap the dynamo body with a rubber or fibre mallet.

- Cover the opening in the cylinder block.

Removing the cylinder head

- Remove the carburettor.

- Remove the exhaust manifold and heat shield.

- Remove the camshaft housing.

- Remove the 4 spark plugs (7720).

- Remove the t cylinder head fixing bolts (7190) or nuts (7259) gradually, in turn. *Take care:* Bolts or nuts that are excessively tight can easily break, resulting in a broken bolt or stud remaining in the block. Try to avoid this by using penetrating oil or heat, as needed.

- Free the cylinder head from the cylinder block. Tap with a rubber or fibre mallet if needed.

- Remove the head gasket (7009) which is normally discarded, or the 4 head sealing rings (9526) which may be re-usable.

- Cover the cylinder block.

Removing the gearbox

In special cases, when *only the gearbox* is to be removed, the following procedure may be used:

- Disconnect all battery leads. Disconnect the earth lead.

- Disconnect the 'I' lead from the ignition coil (see wiring diagram where needed).

- Remove the ignition coil (8054 or 7706). (See *Removing the engine from the frame*).

- Remove the fuel line (7683, 8183 or 8577).

- Remove the exhaust manifold and heat shield (see above).

- Disconnect the gear shift linkage and clutch cable (and pullrod if fitted) from the gearbox. (See *Removing the engine from the frame*).

- Remove the tool box (7610) if fitted under the frame.

- Support the motorcycle by blocks under the footrests and remove the centre stand (7300 or 8328).

- Support the sump on blocks so that the wheels just touch the ground.

- Remove the four bolts (7269) securing the engine in the frame. Ease the supported engine as close to the front mudguard as possible.

- Place rag or a container to contain oil spillage from the gearbox and remove the 6 nuts (7259) which secure the gearbox to the crankcase and sump.

- Withdraw the gearbox from the 6 studs. *Take care* that the oil feed pipe is clear and lift the gearbox from the frame.

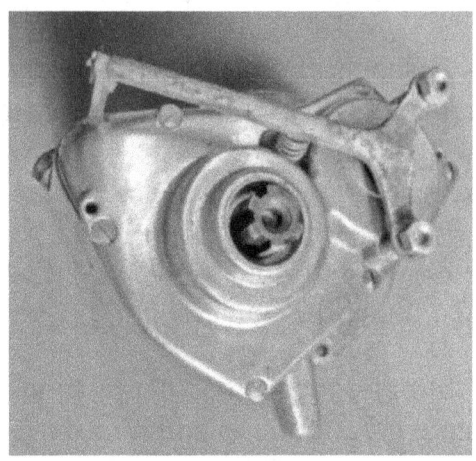

Dismantling gearbox

Once the engine is out of the frame it can be completely dismantled. An engine stand allows the engine to be readily moved or inverted when work is being done on it.

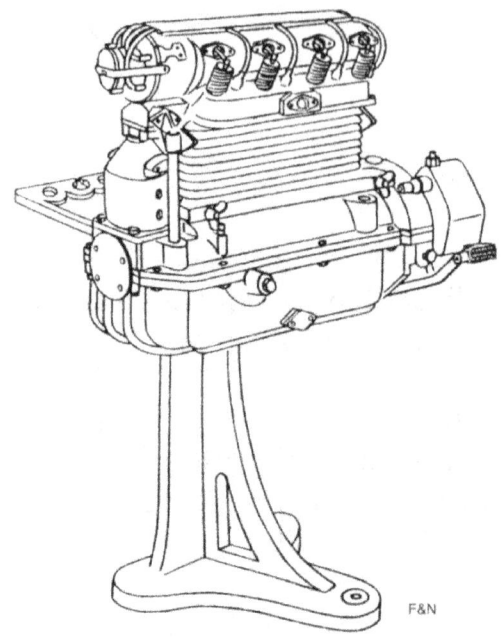

If an engine stand is not available:

To remove the camshaft housing, cylinder head, gearbox or dynamo, rest the engine upright on its sump.

To access the oil pump, crankshaft with flywheel, clutch, connecting rods with pistons, and oil feed pipe are to be removed, invert the engine to rest on its cylinder block gasket surface and remove the sump.

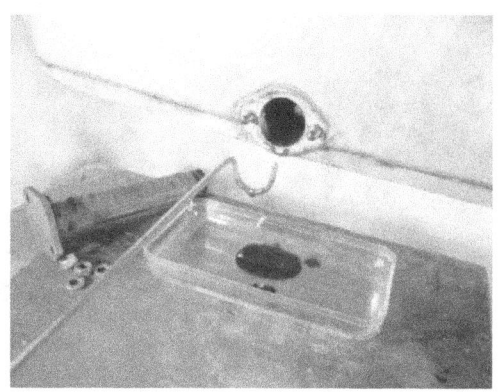

If the engine is inverted with the camshaft housing, cylinder head and dynamo still in place, it is important to support the engine such that it does not rest on the camshaft housing alone.

Removing major components

The carburettor, exhaust manifold and heat shield will already have been removed as part of removing the engine from the frame.

- Drain the lubricating oil. Set the oil strainer (7248) aside and block the oil drain hole with a cloth.

- Removing the camshaft housing: see *Engine – Operations possible with the engine in the frame* 'Removing the camshaft housing'.)

- Removing the dynamo: see *Engine – Operations possible with the engine in the frame*, 'Removing the dynamo')

- Removing the cylinder head: see *Engine – Operations possible with the engine in the frame,* Removing the cylinder head'.)

- Removing the gearbox:
* Remove the 6 nuts (7259) which secure the gear box to the crankcase and sump.

* Withdraw the gear box from the 6 studs. *Take care* that the oil feed pipe is clear.

Note: Although the engine oil has been drained, there will be some oil remaining in the gearbox. Use a rag or container to contain oil spillage.

- Remove the dip stick from the crankcase and invert the engine so that the sump is uppermost.

- Removing the sump:
Take care: Ensure that the oil strainer (7248) has been removed.
 * Remove the 4 bolts (7265) of the crankshaft bevel gear cover plate (7104).

* Remove the cover plate and its gasket (7105).
* Remove the nuts (7259) and spring washers from the 12 bolts (7263 or 7263-2) at the crankcase to sump joint and push the bolts free.
* Drive out the 2 dowels (7565) from the crankcase.

Lift the sump from the crankcase. If the sump is not immediately released, try a sharp tap from a soft faced mallet on the kick-starter shaft.

Removing the oil pump (7320) and oil pump intake pipe (7694):

- Remove the 4 screws (7500) and spring washers (7695) which attach the oil pump intake pipe to the oil pump.

- Remove the screw (7500) and spring washer (7695) which connects the oil feed pipe (7954) to the pump.

- Remove the 4 bolts (7509 or 7191) which secure the oil pump to the crankcase and remove the pump.

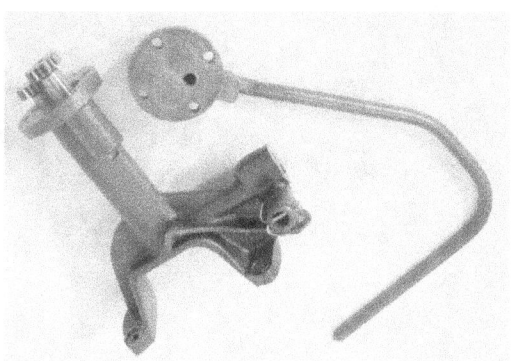

Removing the crankshaft and flywheel, clutch, and kick-starter ratchet assembly:

- Flatten the locking plates (7525) of the 4 nuts (7786) on the main bearing caps.

- Remove the 4 nuts (7786).

- Pull both main bearing caps (7007 og 7008) from the studs.

- Detach the big-end caps from the connecting rods one by one.
Note how the caps and rods are marked, and how the markings are oriented. Caps, shims and rods are *individually matched* and must be reassembled in exactly the original arrangement.

- Keep the individual big-end bolts (7025) in their caps, with their shims (8229) in place.

Note: Use tie-wrap or cable ties to keep items together if the special holders (N 42 or 9021) are not available.

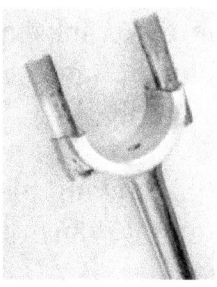

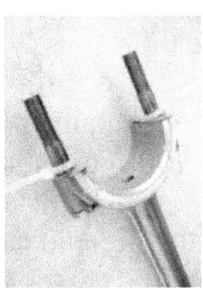

- Lift the crankshaft and flywheel, clutch and kickstarter ratchet assembly free of the crankcase.

- Withdraw the connecting rods, (7020 or 10180), pistons, and rings from the cylinder block. Keep them in order for reassembly.

- Remove the oil feed pipe (7954). Note that this copper pipe fits in a number of slots in the crankcase. It can be freed by gently easing it to and fro from each slot, if necessary using a soft-faced lever.

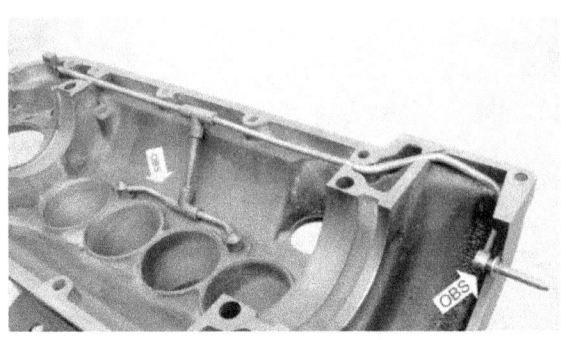

Assembling major components

Fitting the pistons to the connecting rods

Fitting the piston rings

Nimbus-C pistons have either three or four rings. The oil control ring (8732-2, 8733-2 or 8734-2 and 10150, 10151 or 10152) can be identified by the recess machined on its outer edge. It must be fitted in the third ring groove from the top with its recessed face on the underside.

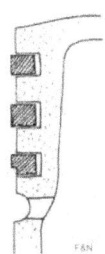

Fitting the pistons onto the connecting rods

It is important that if pistons are being reused, they must be fitted to the <u>same</u> connecting rods and in the same orientation as before disassembly. After fitting to the rods, new small end bushes must be reamed to size with a 16mm reamer.

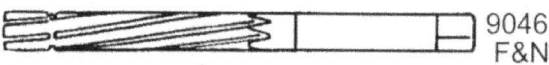

- Fit one circlip (8227) to the piston taking care that it is correctly seated in the groove with its sharp-edged face outward.

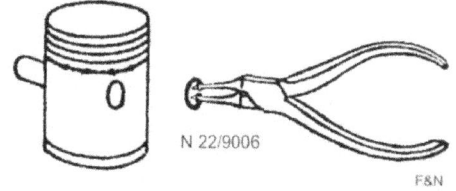

- Press the gudgeon pin (8196) a few millimetres into the other side of the piston.

- Position the piston over the connecting rod (7020 or 10180) so that the gudgeon pin can be pushed through the small end bush. Press the gudgeon pin home so that it contacts the previously fitted circlip.

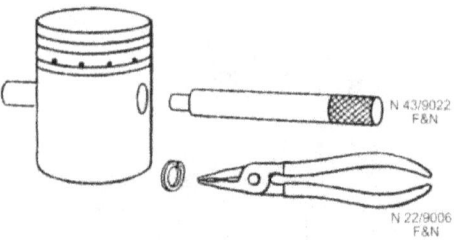

Make sure that the piston can both rotate and slide freely on the gudgeon pin.

Assembling the gear box

- Fit the gear selector shaft bush (7071) if it has been removed. Ensure that the hole for the selector spring and ball aligns with the threaded hole in the gear box housing. If the bush is a replacement, ream it to size with a 16mm reamer.

In case of the 'early' gear box, the selector fork (7069) must be fitted to the gear selector shaft (7819) so that it is rotated 8° from the notch in the shaft (see drawing 7059-3). Fit castellated nut (7238) and lock it with a new split pin (4518)

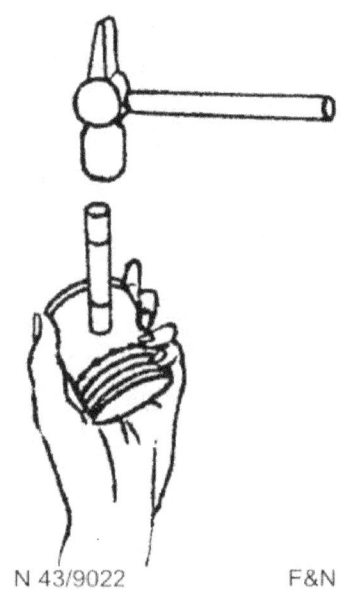

- Fit the other circlip (8227), once again taking care that it is correctly seated in the groove with its sharp-edged face outward.

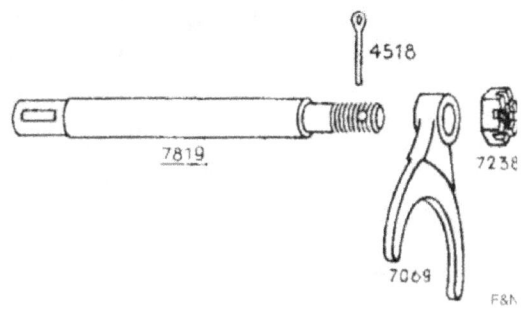

- Fit the two locating dowels (7241) into the gear box housing.

- Fit the two ball bearings (7062/SKF 6305) onto the mainshaft. An oil slinger disc (7049) has to be fitted between 1st gear and the bearing. The disc should be repaired or replaced if it has been deformed or damaged during disassembly. It must be oriented so that there is a 0.5mm clearance between it and the outer bearing race.

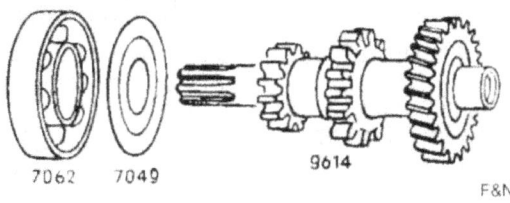

- Fit ball bearing (7062/SKF 6305) to the gear box cover. It should be a press fit and will need to be lightly driven in. If necessary, heat the cover. Apply bearing-grade 'Loctite', or an equivalent product, if the bearing is not a tight fit in the cover.

If the gearbox housing has a layshaft bearing seat (9879), fit it if it has been removed.

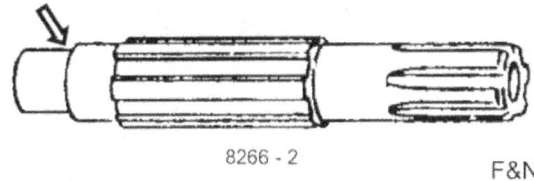

- Fit the two ball bearings (7062/SKF 6305 and 7088/SKF 6304) to the gear box housing. Use open, not sealed bearings. They should be a press fit and will need to be lightly driven in. If necessary, heat the housing. Apply bearing-grade 'Loctite', or an equivalent product, if a bearing is not a tight fit.

- Fit the clutch push rod oil seal (7973-3) to the gearbox cover so that the open side faces inwards - away from the clutch push rod thrust bearing.

- Fit the inner clutch push rod (7781 or 9213). Note that if an inner clutch push rod with captive thrust bearing (9213) is being fitted do not at this stage fit the outer (short) push rod (9215 or 9216).

If the mainshaft 3rd gear has straight-cut teeth, the thrust bearing has to be packed with grease. If the 3rd gear wheel has helical teeth this is not essential as the main shaft is drilled so as to provide oil lubrication to the bearing.

If the inner clutch push rod is part 7781 (with a non-captive bearing) then five new 1/4" balls should be pressed into a greased bearing cup followed by a greased second cup and the outer clutch push rod (7782).

- Fit the lay shaft (8266/8266-2) and 1st, 2nd and 3rd gear wheels (9616 and others), selector fork rod with fork (9210 or 7819 and 7069) and main shaft (7063, 9204, or 9614).

- Put the large 1st gear wheel on the bottom of the gear box housing with the flat side facing away from the bearing.

- Place the selector fork in the 2nd gear wheel with the groove for the selector fork facing away from the end of the gear selector shaft and slide the shaft through the selector shaft bush.

- Slide the lay shaft through the 2nd gear wheel (the shifting gear wheel) and the 1st gear wheel (the largest) and locate the shaft in the lower bearing of the gear box housing. This job requires some patience, but it can be done…

- Push the main shaft fully in place in the gear box housing.

. Fit the 3rd gear wheel to the layshaft, oriented so that the gear dogs can engage.

- - -

Make the following checks before fitting the gearbox cover:

- Is the selector fork correctly placed in the groove of the 2nd gear wheel at the side facing away from the cover and is the gear wheel free to rotate?

- Is the 1st gear wheel (the largest) clearing the oil slinger disc on the main shaft?

- Is the inner clutch push rod correctly placed in the main shaft?

- If the outer (short) clutch push rod has a bearing cup fitted, can the rod be installed now?

- Are both locating dowels correctly placed in their holes?

- - -

- Apply gasket sealant on both the gearbox and cover joint faces and allow it to dry for a few minutes.

- Put the gear box cover in place over the locating dowels and using a soft-faced hammer gently tap the cover into place.

- Fit the seven bolts (7168 or 9203) securing the gearbox cover to the gearbox housing.

- Fit the outer clutch push rod (if this has not yet done – see above!).

- For later gear boxes, press the rubber seal (9220) into the recess provided around the outer clutch push rod.

- On early gearboxes, fit the double-ended shouldered stud (7077) and the dowel-ended stud (7080). - On later gearboxes fit the two double-ended shouldered studs (9219), applying gasket sealant to the threads.

- On early gearboxes fit the clutch release arm (7076) with two springs (7078 and 7081), the adjustment nut (7079), and split pin (3865). On later gearboxes fit the clutch release arm (9218) with two short springs (7078) and adjustment nuts (7079), and a single longer spring (9505).

- Place the 5/16" detent ball (3845) and spring (7074) into the threaded hole above the gear selector shaft and fit the threaded plug (7075 or 7075-2).

Fitting the kickstarter to the sump

Fit the key (7054) to the kickstarter shaft (7108).

Fit the kickstarter return spring (7588) and the spring abutment pin (7563) to the shaft.
Slide the kickstarter quadrant (7109) onto the shaft, shouldered side first, almost up to the key. (Note that in the Nimbus Parts Catalogue the kickstarter quadrant is incorrectly shown <u>reversed</u>).

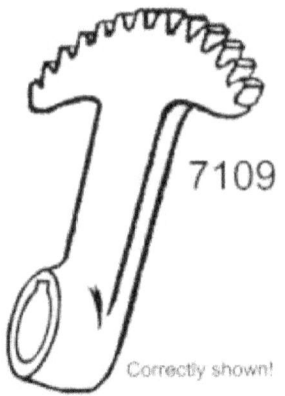

Lubricate the kickstarter shaft and slide it through the kickstarter pivot hole so that the square end of the shaft protrudes from the sump, but the quadrant has <u>not</u> engaged the key.

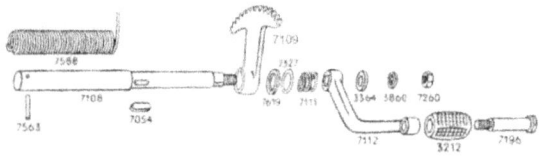

It is now possible to tension the spring by turning the 13.5mm square end of the shaft in the direction of kickstarter pedal travel (Use a good quality adjustable spanner so as to avoid damage to the square). With the quadrant in the 'down' position, turn the kick-starter shaft until the key aligns with keyway in the quadrant and the tension on the spring is sufficient. Then slide the quadrant onto the key using a lever between the quadrant and the sump end.

- Carefully tap the kick-starter shaft home until the shoulder abuts the quadrant.

- Fit the O-ring or conical leather ring (7619) with washer (7327), and spring (7111) onto the outer end of the kick starter shaft.

- Fit the kick-starter crank (7112) with washer (3364), spring washer (3860) and nut (7260).

- Fit the kick-starter pedal (3212) using the special bolt (7196).

Assembling the cylinder head

The cylinder head face must be flat. This can be checked using a steel ruler. A maximum deviation of 0.15mm is allowed. If the head was fitted with four separate gasket rings, refacing the head will result in a change to a single one-piece head gasket being necessary, as the recesses for the gasket rings will have been machined away.

- Fit eight <u>new</u> valve guides (7012). Heat up the cylinder head evenly and cool the valve guides down. Use either the special purpose punch or a soft-faced hammer and hold the cylinder head by hand while working on it.

N 48/9027
F&N

- Fit the valves (10890 or alternatives). Each valve has to be ground into its seat, whether it is being re-used or has been replaced. Before valve-grinding, it is best to have the valve seats cut at a specialist workshop so that they are all at the same depth in the head.

Place the cylinder head face down and arrange some support for the valves, for example a small wooden block placed in the combustion chambers.

- Fit an insulating washer (8194), valve spring seat (7016), inner valve spring (7014), outer spring

(7015) and top collar over each valve guide.

Use a valve spring compressor or a 20 / 22mm ring spanner to press down the top collar down sufficiently to slip two valve collets (8096) around the valve, fully engaging the groove in the valve stem.

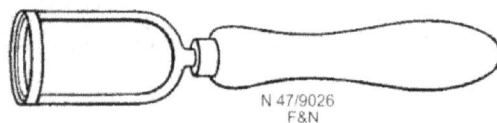

Assembling the camshaft housing

Oil feed pipe and camshaft bushes

If the internal oil feed pipe (7333) has been removed, install a <u>new</u> replacement pipe (7333-2).

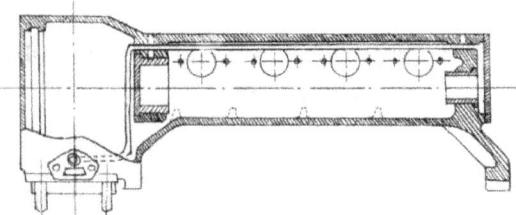

If the front and rear bushes (7469 and 7470) have been removed, install <u>new</u> bushes. Heat up the camshaft housing and drive in the front bush (7469) with the service tool N49/9028, and the rear bush (7470) with the service tool N50/9029. Ream the rear bush with a 20mm reamer and test the fit of the camshaft. If the camshaft cannot be rotated by hand, the two bushes have to be line-bored by a specialist workshop.

Fitting the rockers

- Place the rockers (7096) in the rocker mounts (7097 or 7097-3) noting that oil drain cut-away in the rocker mount is on the underside. Fit the rocker pivots (7098) through the mounts and rockers.

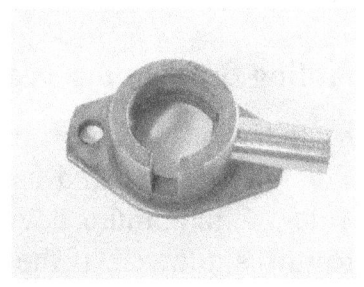

- Fit the eight rocker mounts and rockers to the camshaft housing. Use paper gaskets (7733-2) and gasket sealant.

- If the rockers and mounts are designed for oil seals, fit red silicon rubber seals (10219-1) and metal covers (10230). Fit spring washers under the fasteners and tighten them.

- Fit the camshaft housing end plate (7464) using paper gasket (7520) and gasket sealant. Fit spring washers under the four bolts (7265) and tighten them.

Fitting the camshaft bevel gear

Camshaft bevel gears are either straight cut (7094), skew (7094-2) or helical cut (7094-1).

- Fit the key (7054 or 8424) in the camshaft keyway. If the camshaft bevel gear is straight-cut (7094), and is *not* stamped '2', then use offset key (8424) fitted with the punch mark forward.

- Fit the camshaft bevel gear, using service tool N49/9028 or similar.

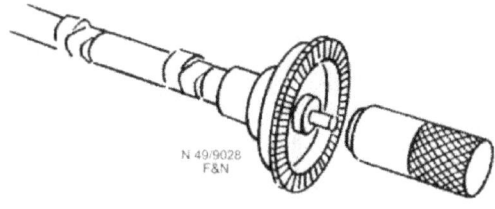

- Fit the two bobweight pivots (7403) to the bevel gear. Note that the gear has five holes. The hole <u>closest</u> to the marked tooth (and closest to the centre) allows lubrication. The hole which is <u>next closest</u> to the marked tooth, is for one bobweight pivot and the second pivot fits in the diametrically opposite hole.

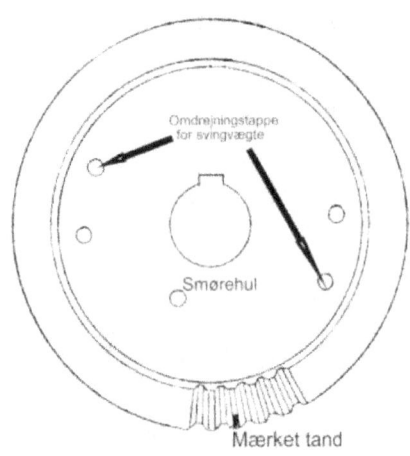

- Fit a pivot bush (7402) to each of the two bobweights (7699) and place the bobweights on the pivots so that the remaining holes in the bevel gear remain visible. (This is essential -see illustration)

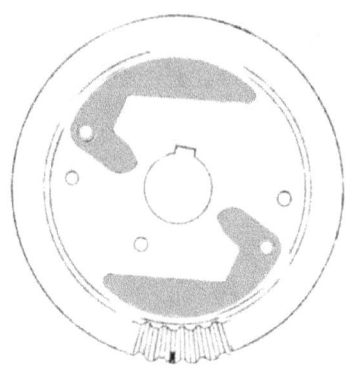

- Fit two spring attachment pins (7404) to base of the contact breaker rotor (7400).

- Fit the contact breaker rotor so that the milled slot on the forward edge is towards the same the lubrication hole and the marked tooth of the bevel gear.

- Fit springs (7582-3) between the attachment pins on the rotor and the bobweight pivot pins.

- Fit the cover plate (8052). It should clear the four-lobed contact breaker rotor by about 0.5mm The posts of the locking plate are riveted at the back of the bevel gear. If the cover plate is being reused, apply a locking product such as 'Loctite'.

Camshaft housing assembly: Completion

- Fit the four studs (7542) which will be used the secure the dynamo sealing collar

- Fit the eight rocker adjusters (7099) and locknuts (7258) to the rockers. Screw the adjusters fully in, but do not tighten the locknuts yet.

- Place the camshaft with bevel gear and contact breaker rotor in the camshaft housing.

Assembling the distributor

The distributor body comes with the following components already riveted in place:

- oil shield (8174)
- condenser clip (7396)
- ignition coil contact bar (8115) and contact bar insulator (7391)
- contact breaker pivot pin (7397),
- eccentric points gap adjuster (7398)
- slotted ignition timing indicator (8121).

Make sure that all rivets are tight.

- Clean the contact breaker points (8053). Points which are being re-used may need to be dressed with fine emery paper and then cleaned with solvent. For new points cleaning with solvent is sufficient.

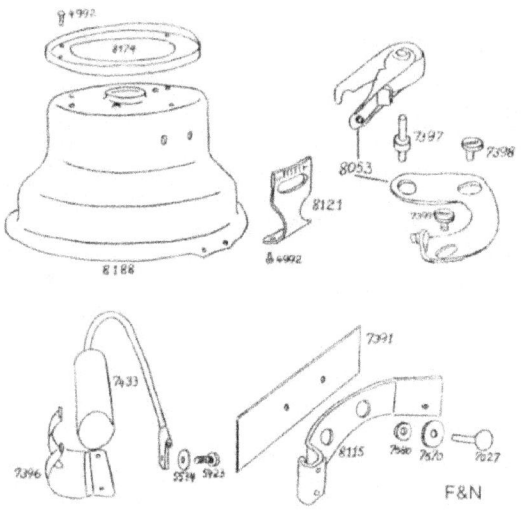

- Place the fixed contact over the contact breaker pivot making sure that the slotted head of the eccentric adjuster (7398) is correctly located in the elongated hole. Fit the adjustment locking screw (7399) but do not yet tighten it.

- Apply a little petroleum jelly ('Vaseline') to the heel of the opening contact.

- Fit the opening contact on the pivot pin and slide the slotted end of the spring under the attachment screw (5423) on the ignition coil contact bar (8115).

- Fit the condenser (7433) in the spring clip. Make sure that the clip is tight, ensuring a sound contact between condenser and earth. Add the condenser lead terminal to the attachment screw (5423) on the ignition coil contact bar (8115) and tighten it with a 6mm open end spanner.

- Check that the contact breaker points align and that the eccentric points gap adjuster (7398) moves the fixed contact as required.

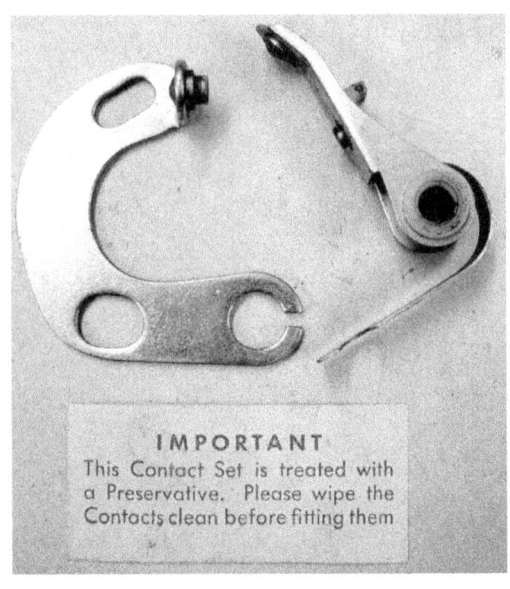

IMPORTANT
This Contact Set is treated with a Preservative. Please wipe the Contacts clean before fitting them

Final Assembly

Assembling the cylinder block

The cylinder block must be thoroughly cleaned before fitting. It is especially important that <u>absolutely no trace</u> of sandblasting material remains in any drilling or crevice.

- Fit the gearbox attachment studs: three plain studs (7004) for an 'early' gear box or three shouldered studs (9219) for the 'later' gear box.

- The cylinder head and camshaft housing may be fixed to the cylinder block on 14 studs (7003) or by 14 bolts (7190The dynamo seal collar may be fixed to the cylinder block on four studs (7005) or by four bolts (7191).

- Fit the two sump locating dowels (7565) into the cylinder block to sump joint face but do not drive them fully home until the sump has been fitted.

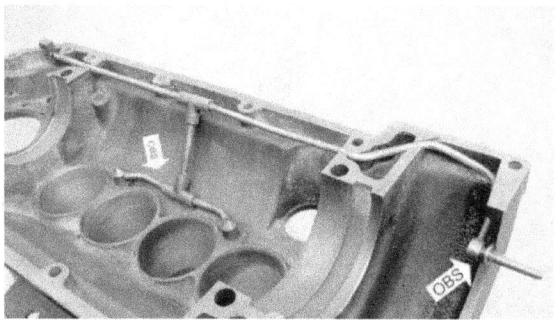

- Fit the oil feed pipe (7954).

Stand the cylinder block on the block to head joint face, protecting the studs (if fitted). Carefully press the oil feed pipe into the slots and holes in the crankcase. Take care to ensure that the two jets close to the bores are pressed fully home in the holes provided and that the collar on the feed pipe to the gearbox fits precisely. This may require slightly bending or adjusting the feed pipe.

- Fit the four main bearing cap studs (7006). These studs have one flat end, which has to be screwed home in the cylinder block. To tighten a stud, fit two nuts (7886) on the outer threaded end, lock them to each other, and use a ring spanner.

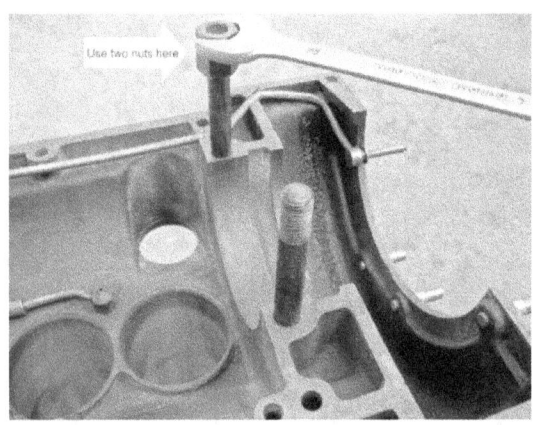

Installing assembled connecting rods and pistons

- The marked side of the connecting rod normally faces away from the oil feed pipe. Check that both circlips (8227) are correctly fitted.

- Piston ring gaps should be staggered - not aligned.

- Lightly lubricate both cylinder bores and pistons.

- Insert the pistons in their bores.

Fitting the crankshaft main bearings

Main bearings (7052) are SKF6407-C2 type (open deep groove ball bearings with reduced internal clearance).

- Clamp the crankshaft in a vice close to the bearing which is to be fitted.

- Heat the bearing, if necessary.

- Drive the bearing into place, using special tool N45/9024 or a drift made from pipe with an inner diameter of 35mm and accurately squared-off ends.

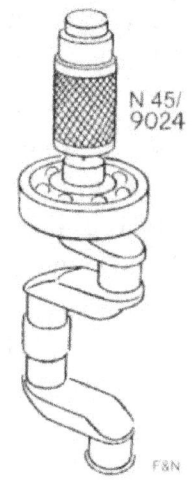 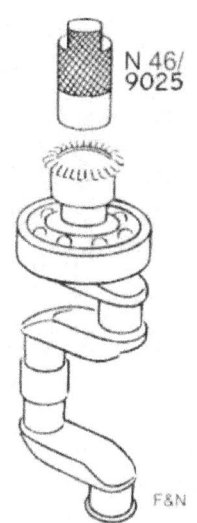

Fitting the crankshaft bevel gear

Original bevel gear (7053)

- Heat the bevel gear if necessary.

- Insert the same number and type of shims (7577 or 8274) between the main bearing and the crankshaft bevel gear as were present on disassembly, or if a new bearing or bevel gear is to be fitted, insert one 0.15mm shim (7577).

- Fit the key (7054).

- Drive the bevel gear home, using special tool N46/9025 or equivalent

Helical-cut bevel gear (7053-1)

- Heat the bevel gear if necessary.

- Fit the key (7054)

- Drive the bevel gear home with a drift cut from pipe with an inner diameter not less than 35.3mm and an outer diameter of not more than 40mm. Both ends should be squared off, and a 6.5mm x 3mm notch cut in one end to clear the key. Without this notch the key will be damaged. Special tool N46/9025 should not be used - it too will damage the key.

Adjusting the lower bevel gear backlash

It is good practice to make sure that the clearance between the crankshaft and dynamo bevel gears is correct before the crankshaft, flywheel, and clutch are finally installed.

- Dry mount the assembled dynamo (without gasket sealant) on the cylinder block.

- Put the crankshaft with fitted main bearings and flywheel in place in the cylinder block.

- Tighten the four main bearing cap nuts, leaving out the square lock washers.

- Turn the crankshaft and observe the backlash, which should be about 0.20mm. Adjustment washers can be placed between the crankshaft bevel gear and the main bearing until there is just zero clearance, and then shims totalling 0.15mm to 0.20mm should be removed. With worn gears it may be necessary to accept a greater clearance of 0.30mm to 0.40mm to reduce noise with the clutch disengaged.

Fitting the kickstarter ratchet assembly and flywheel

Early type kickstarter ratchet:

- Fit the spring for the ratchet disc (7586).

- Fit the ratchet disc (7473).

- Fit the ratchet wheel (7036).

If the ratchet disc or the ratchet wheel is worn, it may be possible to recondition them with a grinder. Otherwise the system has to be converted to the later type. This requires modifying the flywheel. This operation is described in Fisker and Nielsen's *Technical Circular No. 75, September 1953* and can be carried out by a specialist Nimbus workshop.

Later type kick-starter ratchet:

- Fit each of the two pawls (9833) to the pawl pivot bolts (9834) and fit the bolts in the holes through the flywheel, orienting the pawls as shown.

- At the clutch side of the flywheel fit the dual lock washer (9836) over both pawl bolts, fit the nuts (7260) and lock both.

- Fit a spring (9835) in each pawl.

- Fit the ratchet gear wheel (9832).

Fitting the flywheel to the crankshaft

The crankshaft and flywheel tapers must be completely clean and burr-free.

- Place the flywheel with the assembled kick-starter ratchet on a solid surface.

- Fit the key (7564) into the key way in the crankshaft taper.

- Fit the crankshaft taper into the flywheel and test the kick-starter ratchet.

- Clamp the crankshaft in a vice close to the kick-starter assembly.

- Fit the flywheel lock washer and the flywheel retaining nut (7035). Tighten the nut firmly and lock it.

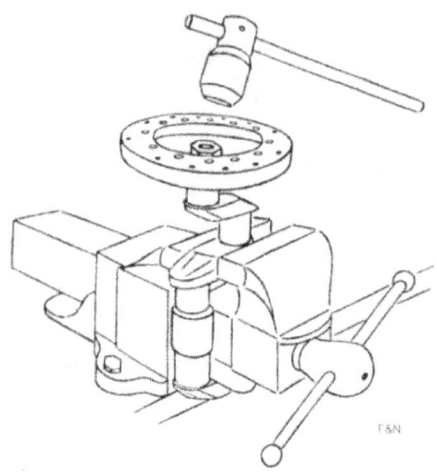

Fitting the clutch to the flywheel

- Clamp the crankshaft in a vice close to the flywheel.

- Insert a bush (7039) in each of the twelve threaded holes around the flywheel.

- Insert a spring (7038-2) in each of the plain holes (between the threaded holes).

- Place the pressure plate (7143–2, 7143-3, or 9592) onto the springs so that the 12 holes in the plate are over the threaded holes.

- Lay the clutch (friction) plate (7763, 8364 or 9536) over the pressure plate with the clutch hub uppermost.

- Add the clutch cover plate (7044–2 or 9593) and carefully centre all three plates. The special clutch centering tool N41/9020 is designed to ensure this.

- Compress the clutch assembly using the three special clutch spring compression tools N44/9023. M8 bolts can be used for this purpose. Make sure that all bushes enter the holes in the pressure plate as the assembly is compressed.

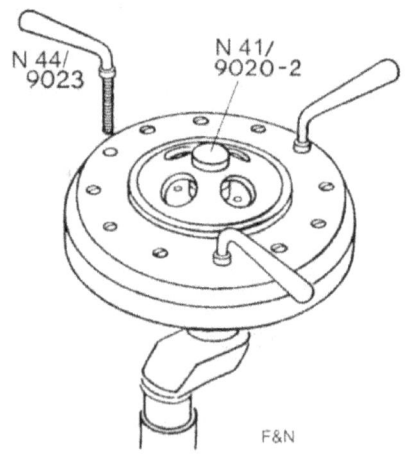

- Fit 9 of the clutch assembly screws (7040), remove the compression tools and fit the remaining three screws (7040).

- Secure the clutch assembly screws with split pins, or by wiring together adjacent pairs of screws ...or

- Tighten the screws with an impact screwdriver.

Fitting the connecting rods and crankshaft

- Stand the cylinder block on the cylinder to head joint face. Use a wooden board (as a substitute cylinder head) or wooden blocks both to protect the head studs and to prevent pistons from being pushed out of the cylinders. Push the assembled pistons and connecting rods, without big end bearing caps, but with bolts and shims retained by the big end nuts, down into the cylinders as far as possible. It is essential to avoid pushing the pistons beyond the cylinder to head joint face.

- Check that the connecting rods and the pistons are in the correct cylinders, and that the identifying stamping marks are away from the oil feed pipe side.

- Lay the assembled crankshaft in position. Make sure that the main bearings are fully in place and check that the flywheel and clutch clear the oil feed pipe.

- Fit the front main bearing cap (7007) which is flanged so as to locate the crankshaft, and the rear main bearing cap (7008) which is plain. The caps are stamped '1' and '2' and there are corresponding stampings on the cylinder block. Fit the four lock washers (7525) and main bearing cap nuts, tighten the nuts but do not yet lock them.

- Pull the four connecting rods up into place on the crankshaft journals.

- Fit the four connecting rods to the journals. Note that the connecting rods and big end bearing caps are marked, and that these marks must match and must be on the same side of the journal. Make sure that the shims are correctly placed.

- Use <u>new</u> spring washers under the big end nuts (7026). After fitting each big end cap, check that the crankshaft can still be rotated freely.

- When all four connecting rods have been fitted, check that the crankshaft rotates. The only friction present should be that between the piston rings and the cylinder bores.

- Tighten the big end bearing cap nuts with a torque of 30-35ft/lbs (41-47 Nm) and the main bearing cap nuts to 45ft/lbs (60 Nm).

- Bend up the square lock washers to lock the main bearing cap nuts.

Note that this phase of engine assembly is most critical. It is essential to give sufficient time to each task, to check frequently when proceeding, and if in doubt, to repeat an operation until satisfied.

Fitting the oil pump

- Fit the oil pressure relief valve to the oil pump housing (7799). The valve consists of a 5/16" steel ball, kept in place by a compression spring (8045) and split pin (4518).

- Fit the two oil pump gears, long (7717) and short (7718).

- Fit the oil pump to the cylinder block using the four bolts (7191) and spring washers, with the pressure relief valve on the same side of the engine as the main engine and gearbox oil feed pipe.

- Attach the main engine and gearbox oil feed pipe to the oil pump outlet using an 5M x 13mm screw (7500) and spring washer.

- Fit the oil pump cover with scavenge pipe (7694) using four 5M x 13mm screws (7500).

- Check that all fasteners are correctly tightened before fitting the

sump. Confirm that the crankshaft rotates without friction other than that from piston rings in the cylinder bores, and check that the flywheel and clutch clear the main oil feed pipe.

Fitting the sump to the cylinder block.

First ensure, by resting the sump on the block, that nothing prevents it fitting correctly. In particular, check that the oil pump scavenge pipe allows the mesh oil strainer to be fitted without distortion. Realign the pipe by hand if necessary.

- Clean the cylinder block and sump joint faces and apply gasket sealant, allowing it to dry for a few minutes.

- Replace the sump, making use of the two locating dowels (7565).

- Fit the 12 screws (7263-2), plain washers, spring washers (3861), and nuts (7259).

- Tap the two locating dowels fully home.

- Tighten all nuts to 20–25ft/lbs (27-34 Nm). Work from the middle, crosswise.

- Fit the mesh oil strainer (7248) using a new cork gasket (7312).

- Fit the bevel gear cover plate (7104) and gasket (7105) using the four M6 x 12mm bolts and spring washers. Apply a little gasket sealant to all contact areas.

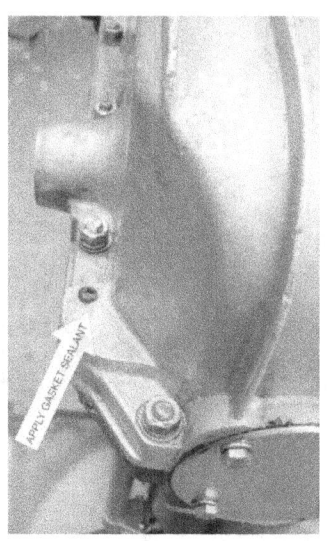

- Insert the dipstick (7242).

- Plug both the oil filler and oil drain holes with clean rag pending further assembly work.

Fitting the assembled dynamo

Apply a little gasket sealant to the joint faces and make sure that the drive spade of the dynamo engages with the slot in the oil pump gear shaft. The marked tooth of the upper bevel gear (8056-2) should align with the mark on the dynamo upper end bracket (7879) ensuring that the same teeth mesh.

Fitting the assembled cylinder head

If the cylinder head fixing is by studs (7003), these must be fitted to the cylinder block. The joint faces of both head and block must be clean and some gasket sealant may be applied.

- Place the cylinder head gasket (7009) on the cylinder block with the seam face uppermost. The original gasket is stamped with an ' I ', which must be on the upper face and to the right for correct installation.

Gaskets from present days are symmetrical.

- Lay the cylinder head in place on the gasket, with the exhaust ports on the right and with the inlet manifold on the left of the engine.

- If the cylinder head is secured by bolts, fit ten new cylinder head bolts (7190) with plain washers. If secured on studs, fit ten nuts (7259) with plain washers.

- Tighten all fasteners, crosswise, working from the centre pair. Torque settings are 30ft/lbs (41 Nm) for nuts and for bolts with head marking '10.9', and 25ft/lbs (34Nm) for bolts marked '8.8'.

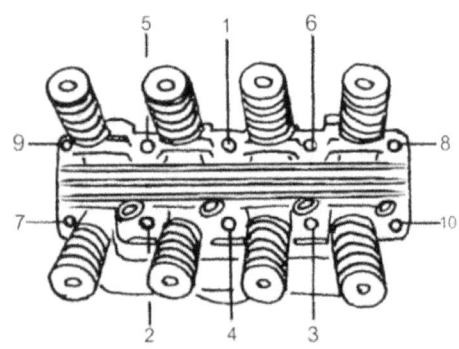

Fitting the assembled camshaft housing

- Turn the engine by using the kick-starter until the adjustment mark on the flywheel can be seen in the centre of the flywheel access hole.

- Place the dynamo upper cover seal clamp (7463) and seal (7516-2) on the dynamo upper cover ('dynamo neck'). Apply some sealant silicone on the inner side of the seal clamp and the seal.

- Turn the camshaft until the marked tooth of the bevel gear relates to the index mark inside the camshaft housing according to the type of gear.

There are three bevel gear types:

* <u>skew</u> cut (7094): straight, angled teeth. The marked tooth must be *one* tooth *before* the index mark.

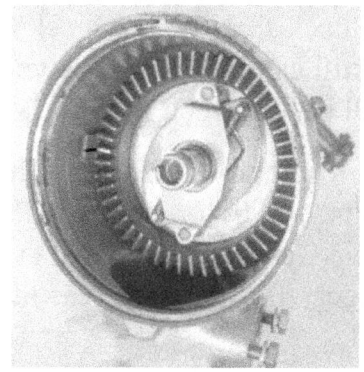

* <u>helical</u> cut (7094-1): curved teeth. The marked tooth must be *one* tooth *after* the index mark.

* <u>straight</u> cut (7094-2): straight radial teeth. The marked tooth must be aligned with the index mark.

- As the camshaft housing is lowered into place, the skew cut and helical cut gears will turn as they engage with the teeth of the dynamo bevel pinion, and both - when fitted - will have the marked tooth aligned with the index mark.

- Check that the flywheel 'I' mark is still central - if not, valve timing will be incorrect.

- Fit the four camshaft housing bolts (7190) or nuts (7258) with plain washers and tighten evenly in stages. Tighten both rear fasteners last.

- Check the bevel gear backlash. First, install shims under the upper dynamo bevel gear until all clearance has gone. Then remove shims totalling 0.15-0.20mm. If there is no clearance to start with, even without shims, check that the front camshaft bush bearing (7469) is fully seated. If that is the case and there is still no clearance, 0.20mm must be machined from the face of the front camshaft bush.

- Slide the dynamo upper cover seal (7516-2) and clamp (7463) up into place and fit the four nuts (7258). Tighten them in stages so as compress the seal evenly and not to deform the clamp and ensure that the clamp does not contact the camshaft housing.

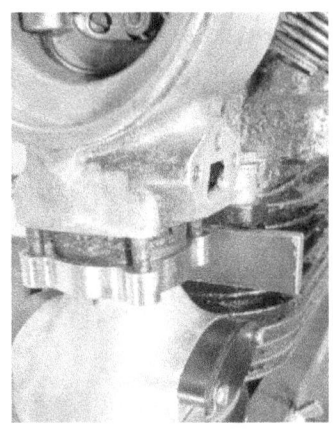

- Fit the camshaft oil feed and drain assembly (7865) using the two screws (7191), spring washers (3862), and the paper gasket (7517) <u>without</u> gasket sealant. If necessary, face both joint surfaces using grinding paste and some flat glass.

Fitting the high tension leads and spark plugs

- Fit the high tension leads through the rubber sleeves (9506 and 9507) in the support brackets (9508 and 9509).

- Fit the support brackets under the front securing bolt (7191) or nut (7258) of the inlet side rocker mountings of cylinders 1 and 3.

- Fit the 4 sparking plugs (7720).

- Fit the plug caps (7510) with terminal (7511 or 9330) onto the sparking plugs.

Fitting the assembled distributor

- Fit the cork gasket (7743) into the circumferential groove in the camshaft housing with the join uppermost, using some non-hardening gasket sealant.

- Fit the distributor, holding the opening contact breaker point clear of the contact breaker cam.

- Turn the engine over by the kickstarter until one of the four contact breaker cam lobes is positioned exactly under the cam follower of the opening contact. Adjust the points gap to 0.7mm by means of the eccentric adjuster (7398) below the cam, measured by feeler gauge. Tighten the fixing screw (7399) above the cam.

Put the rotor (7432) in place, making sure that it fits into the notch on the cam and is pressed fully home. *Do not fit the ignition coil until the engine is in the frame.*

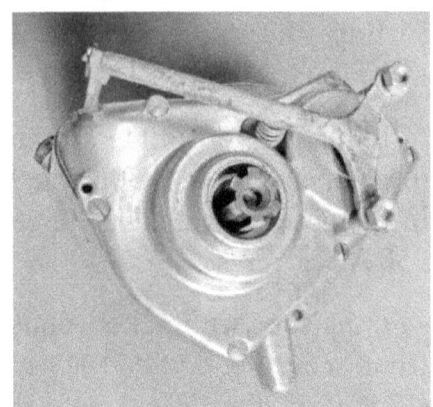

Fitting the assembled gearbox

- Fit the two cork gaskets (7611 for the oil feed pipe and 7612 for the oil return), using gasket sealant.

- Offer up the gearbox, turning the main shaft as needed to engage with the internal splines of the clutch hub, at the same time ensuring that the oil feed pipe is entering the gearbox without any force being needed.

- Secure the gearbox to the studs fitted to the cylinder block and sump using eight plain washers (7206), spring washers (3861) and nuts (7259). Tighten in stages.

- Turn the engine over with the kick-starter and check that all three gears can be engaged. Clutch operation can be checked as well, though this operation may require improvising an extension to the release arm.

- On later type gearboxes (with a 'T-shaped' clutch release arm), pull the release arm in and wire it temporarily in this position.

Fitting the assembled Carburetor

The carburettor should be fitted only after the engine has been fitted into the frame.

- Fit these parts to the intake manifold:

* <u>For early</u> carburettor types 1934-1 and 1934-2, use two bolts (7429) to attach:
- paper gasket (7521)
- carburettor body (7730)
- two spacers (7417)
- choke plate (7497), spring (7583) and intake disc (7501

* For all later type carburettors types use two screws (8594) to attach:
- paper gasket (7521)
- carburettor body (8551)
- choke assembly (8597)

- For carburettor types 1938 and later:
Attach the air filter (8584) by means of three screws (8578). Fit the air filter with the drain hole down.

- Fit the fuel line (7683, 8183 or 8577).

- Fit the crankcase breather assembly (7684, 8583 or 8583-2).

- Pass the throttle cable (7673, 8585 or 8927) behind the petrol tap and fit it to the hook (7211) of the throttle twist grip. Adjust the throttle cable adjuster (7414-2) at the mixing chamber top so that there is a little free play when the throttle twist grip is closed.

Fitting the exhaust

The exhaust manifold and heat shield should be fitted only after the engine has been fitted into the frame. Three different sets of fasteners have been used. In every case apply copper grease or a similar anti-sieze compound to the all manifold fastener threads.

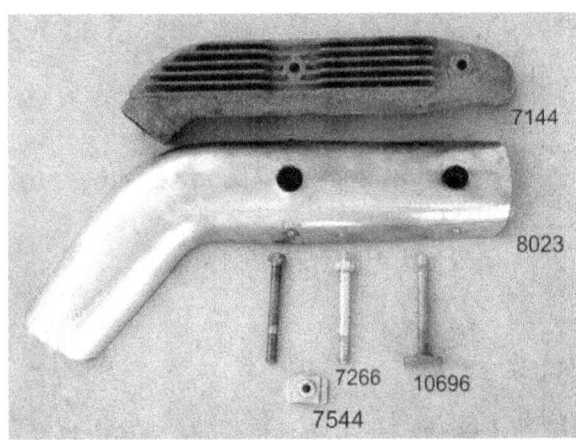

- Place the T-nuts (7544/10696) or T-bolts (10696) between the exhaust ports of cylinders 1 and 2, and cylinders 3 and 4.

- Place the 4 exhaust gaskets (7645) into the exhaust ports.

Offer up the manifold.
- If the fasteners used are standard M8-1.25 x 80mm bolts, or T-bolts (10696), the heat shield (8023 or

8454) must be fitted together with the manifold, using standard 8mm nuts (7259) on the T-bolts.

- If the fasteners are hexagon-waisted studs (7266), first fit and tighten up the manifold only. Then attach the heat shield (8023 or 8454) to the studs by 6mm nuts (7258).

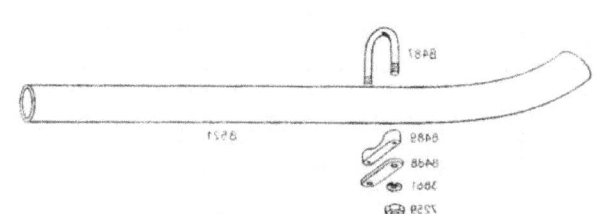

and two 8mm spring washers (3861) and nuts (7259).

- If the exhaust pipe (8073 or 8521) has been taken off, fit it to the frame.

* The low-level exhaust pipe (8073), which includes a baffle and an attached fish-tail, has a welded-on attachment plate which is secured to the lower frame member by two 8M x 19mm bolts, washers and nuts. (The baffle was originally packed with asbestos fibre, but glass fibre can now be used).

* The high-level pipe (also with baffle and fishtail) is attached to the upper frame member frame by a U-bolt, cast aluminium support bracket (8489), flat plate (8488),

Engine repair

Many repairs nowadays have to be left to a professional workshop, as mistakes can become very expensive and irreplaceable original parts may be damaged.
But take a photo copy of the checklists in the next pages as your follow up with the repair.

CYLINDER BLOCK	Cylinder block no.	
	Year:	
	Date:	Sign.

1 2 3 4

- cooling fins (condition) _____
- upper face (flatness) _____
- outer surface
    - sandblasted _____
    - painted _____
- threaded holes
    - upper face (12) _____
    - dynamo (4) _____
    - gearbox (3) _____
- cylinders
    - hone _____
    - bore _____
    - sleeve fitting _____

Cylinder dimensions
(20mm from top)

	Standard	
Standard	60.0mm	acceptable < 60.27mm
1st oversize	60.6mm	acceptable < 60.87mm
2nd oversize	61.2mm	acceptable < 61.47mm
3rd oversize	61.8mm	sleeve if > 62.07mm

- at disassembling				mm Cylinder no.	- at assembling			
1	2	3	4		1	2	3	4
				longitudinal				
				transverse				

Checklist:
- bearing caps _____
- inside painting _____
- oil pipe cleaning _____
- oil pump: gear wheels _____
- oil pump: 5/16" ball _____
- oil pump: spring _____
- oil pump: inlet pipe _____

Knud Jørgensen: NIMBUS - Maintenance. 2012

Cylinder block
Make a copy of the checklist 'Cylinder block'.

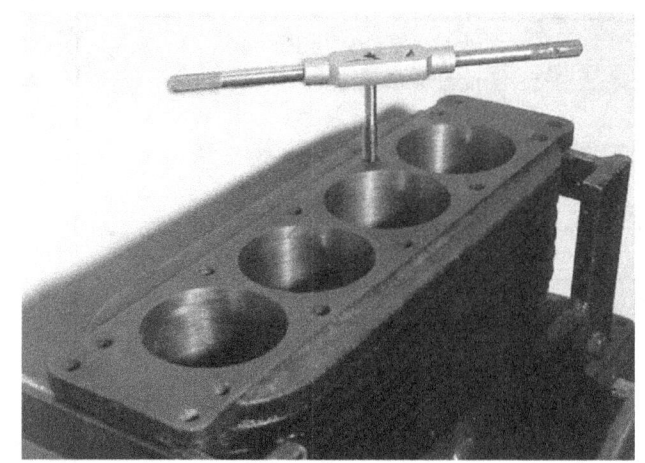

CYLINDER HEAD

Engine no.	
Year:	
Date:	Sign.

- cooling fins (condition) ____
- cyl. block joint face (flatness) ____
- outer surface:
    - sand blasted ____
    - painted ____
- tapped holes (2) ____

Valve seats

	Inl. no.				Ehx. no.			
	1	2	3	4	5	6	7	8
Ground								
Milled 45°								
New seats fitted?								

Valves

	Inl. no.				Ehx. no.			
	1	2	3	4	5	6	7	8
Seats 45°								
Stem diam.								

Standard stem: 7.0mm + 0.02mm / − 0.61mm   Acceptable: 6.9mm   Replace at: 6.7mm

Valve guides

	Inl. no.				Ehx. no.			
	1	2	3	4	5	6	7	8
Inner diam.								

Standard: 7.0mm + 0.022mm / − 0.000mm   Acceptable: 7.2mm   Replace at: 7.5mm

Valve springs

	Inl. no.				Ehx. no.			
	1	2	3	4	5	6	7	8
Inner springs								
Outher springs								

Inner springs: 2.5 kg (5 lb/8.2 oz) = 29mm   4.5 kg (9 lb/14.7 oz) = 23mm
Outher   − : 6.5 kg (14 lb/5.3 oz) = 33mm   11 kg (24 lb/4.01 oz) = 27mm

Knud Jørgensen: NIMBUS - Maintenance. 2012

Cylinder Head
Make a copy of the checklist 'Cylinder Head'.

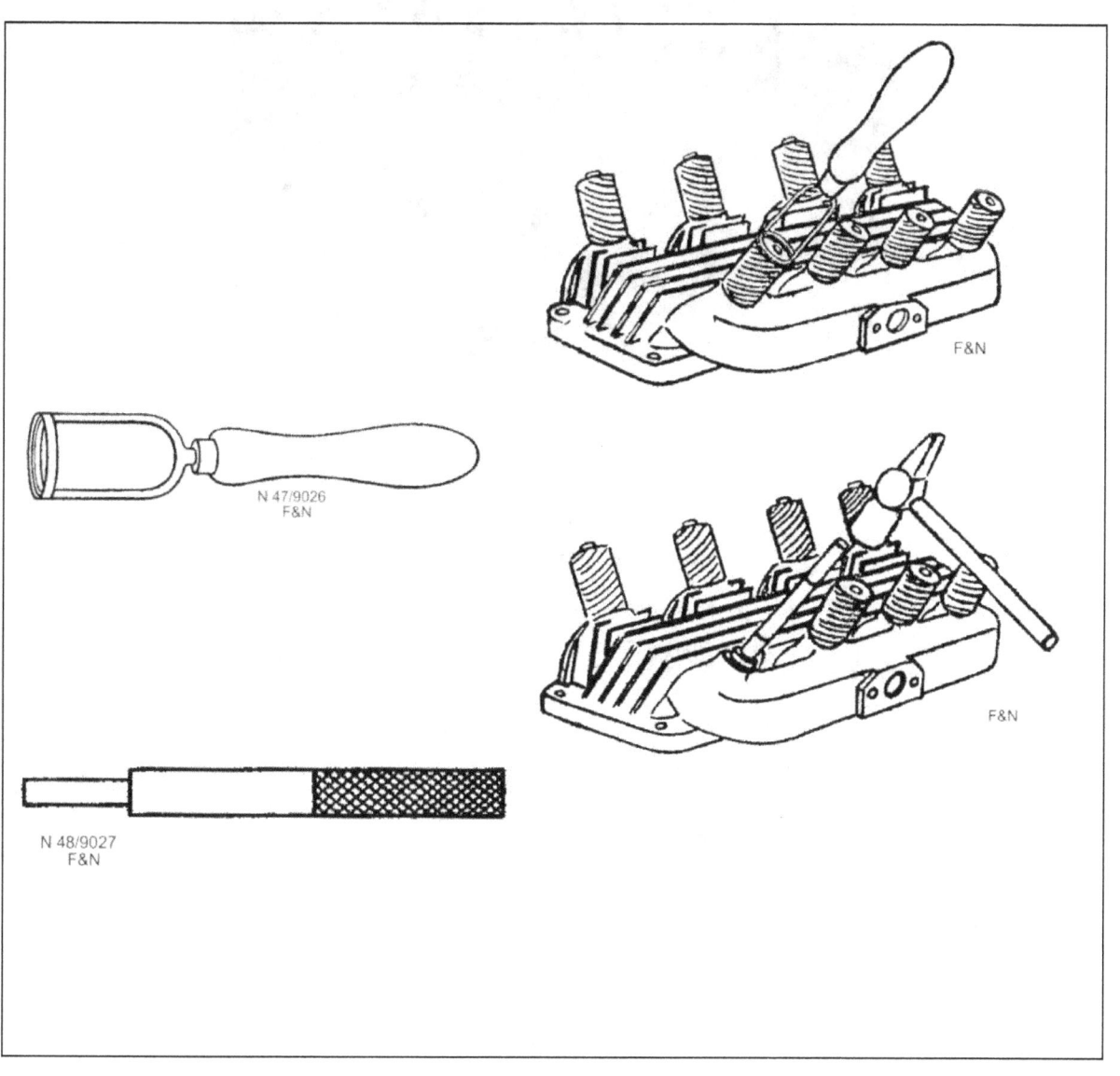

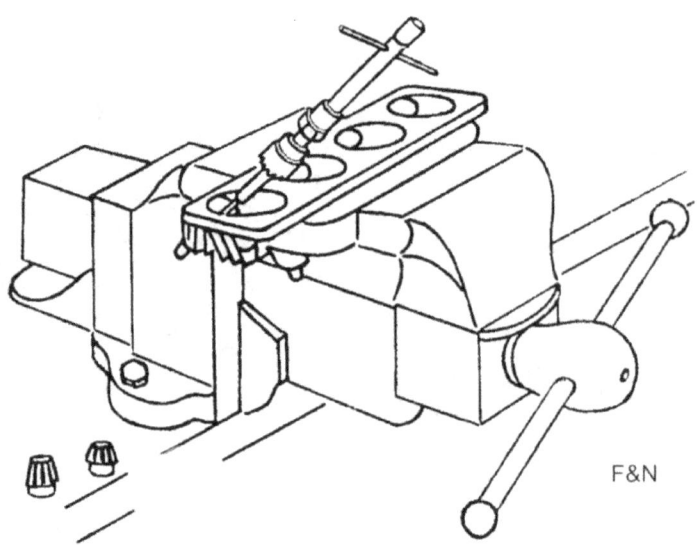

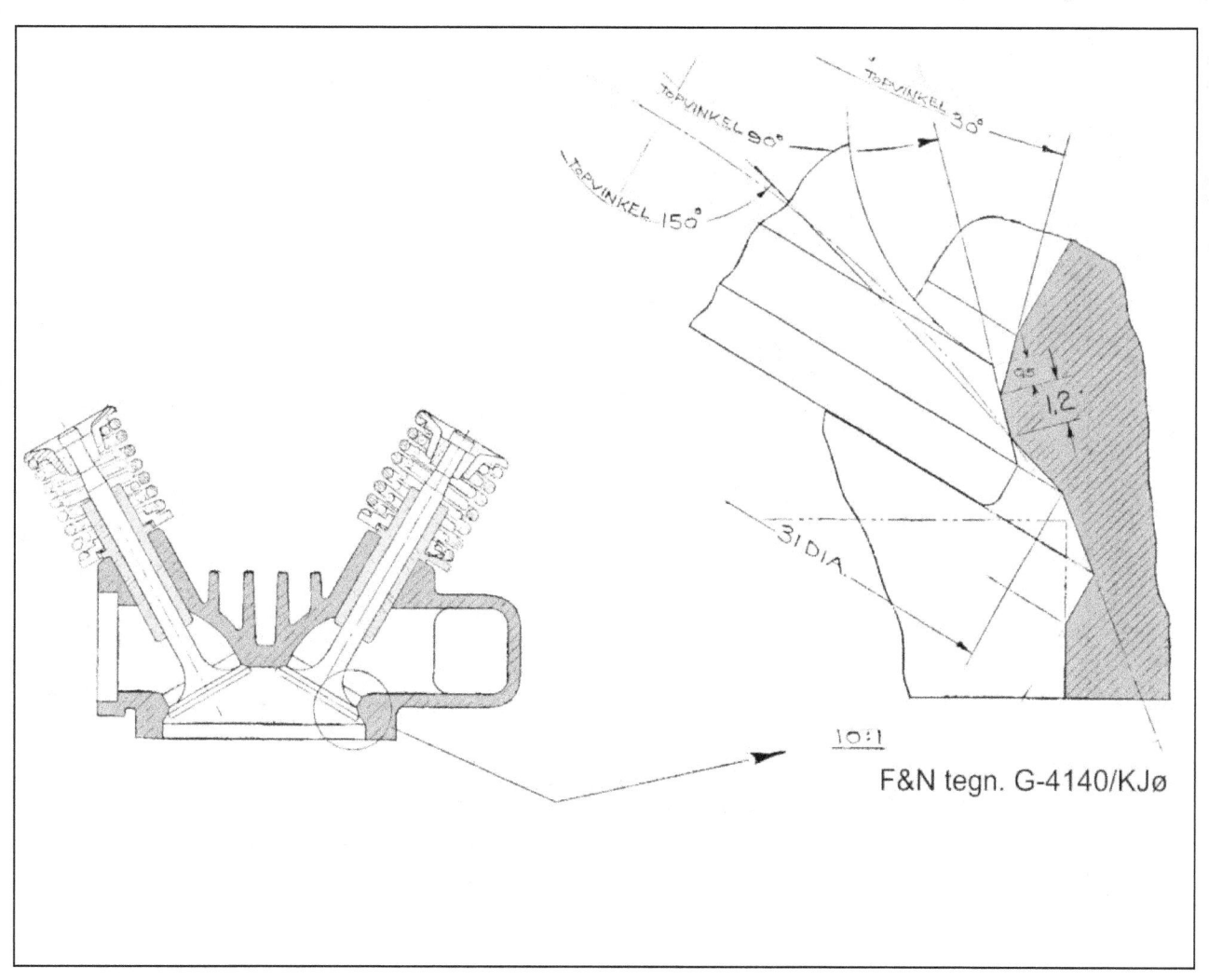

CRANKSHAFT	Engine no.	
	Year:	
	Date:	Sign.

- main bearings
  - front _____
  - rear _____
- shims _____ mm
- thread rear _____
- cleaning oil passages _____
- cleaning key ways _____

Crankshaft journals

		x - x	y - y	ovality
No. 1	a			
	b			
	c			
No. 2	a			
	b			
	c			
No. 3	a			
	b			
	c			
No. 4	a			
	b			
	c			

Standard: 40.0mm + 0.009mm − 0.025mm

Acceptable ovality: 0.03mm. Grinding at 0.08mm ovality.

Crankshaft (assembly):
(Mark with X)

40.0mm ____ 39.75mm ____ 39.5mm ____ 39.25mm ____ 39.0mm ____
38.75mm ____ 38.5mm ____ 38.25mm ____ 38.0mm ____

Crankshaft

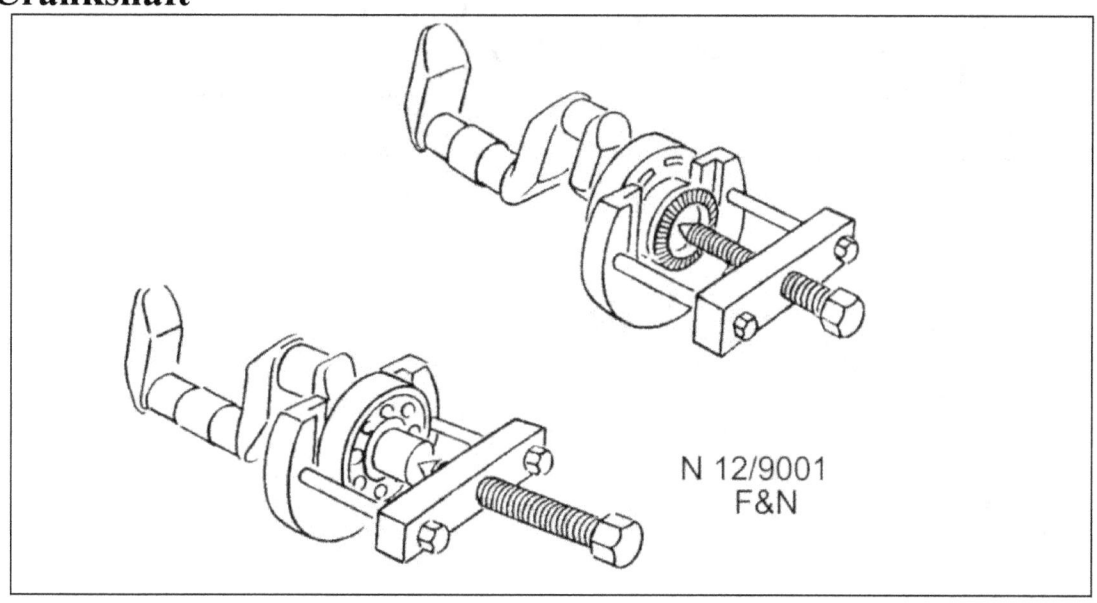

N 12/9001
F&N

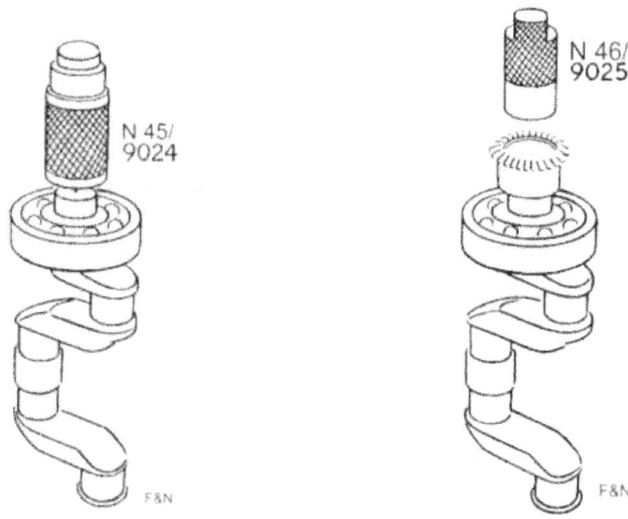

N 45/9024

N 46/9025

PISTONS	Engine no.	
	Year:	
	Date:	Sign.

- type of pistons:
  - 4-rings _____
  - 3 rings _____
  - dome top _____
  - flat top _____
- marked size: _____ mm
- marked clearance _____ mm

Diameters		No. 1	No. 2	No. 3	No.4
top (crown) = ø	ø (x – x)				
bottom (skirt) = n	n (x – x)				
	ø (y – y)				
	n (y – y)				
Max. ovality: 0.12mm	ovality ø				
	ovality n				

Piston rings		No. 1	No. 2	No. 3	No.4
Gaps measured 20mm down bore	top ring				
Standard: 0.25mm - 0.40mm	2nd ring				
Acceptable: 1.0mm	3rd ring				
(if present)	4th ring				

Clearance		No. 1	No. 2	No. 3	No.4
- between ring and groove Max. 0.067mm	top groove				
Max. 0.052mm	2nd groove				
Max. o.072mm	3rd groove				
Max. 0.067mm	4th groove				

Gudgeon pins/piston pins:	No. 1	No.2	No. 3	No.4
Standard: 16.0mm + 0.000mm - 0.008mm	Minimum: 15.9mm			

Knud Jørgensen: NIMBUS - Maintenance. 2012

Pistons
Make a copy of the checklist 'Pistons'.

A piston ring should be free to turn in its groove.

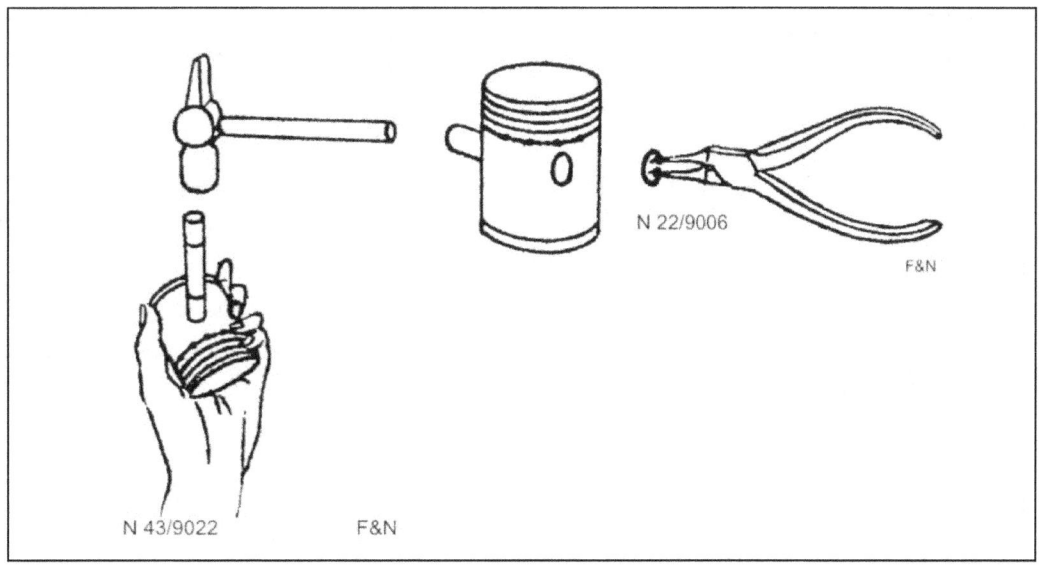

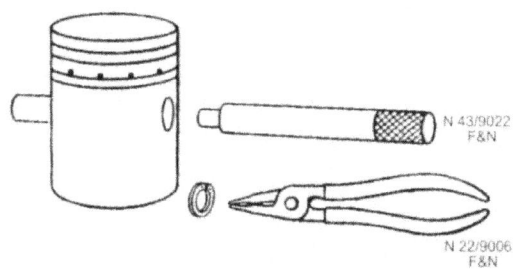

	Engine no.	
# CAMSHAFT	Year:	
	Date:	Sign.

springs _____

Gear wheel type:
- straight _____
- skew _____
- helical cut _____

If skew and NOT marked '2'
special (double) key? _____
auto advance bobweights _____
bushes _____

Cam profile

	No.1		No.2		No.3		No.4	
	Inl.	Exh.	Inl.	Exh.	Inl.	Exh.	Inl.	Exh.
Ø								
H								
h = H - Ø								

Acceptable: h = 5.15mm - 5.2mm. Otherwise: grind!

Rocker arms

	No.1		No.2		No.3		No.4	
	Inl.	Exh.	Inl.	Exh.	Inl.	Exh.	Inl.	Exh.
Rocker arm face								
Adjustment screw								
Rocker arm shaft								
Rubber/silicone gasket								

Check
- camshaft front bearing surface dia. = 35.0mm _____
- camshaft rear bearing surface dia. = 20.0mm _____
- front bush _____
- rear bush _____

Knud Jørgensen: NIMBUS - Maintenance. 2012

Camshaft
Make a copy of the checklist 'Camshaft'.

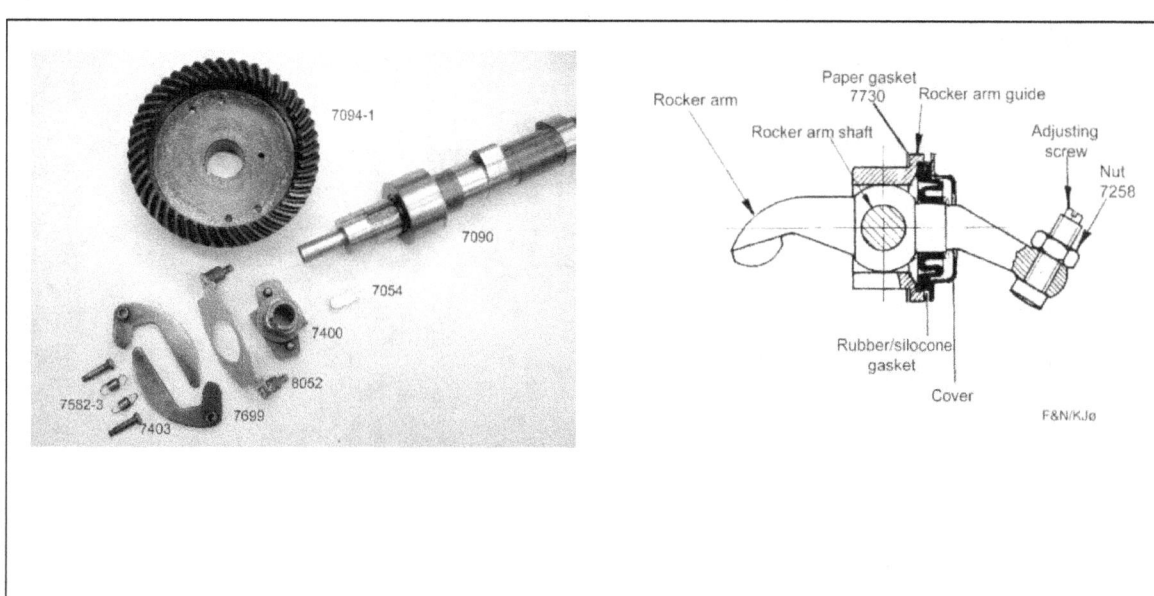

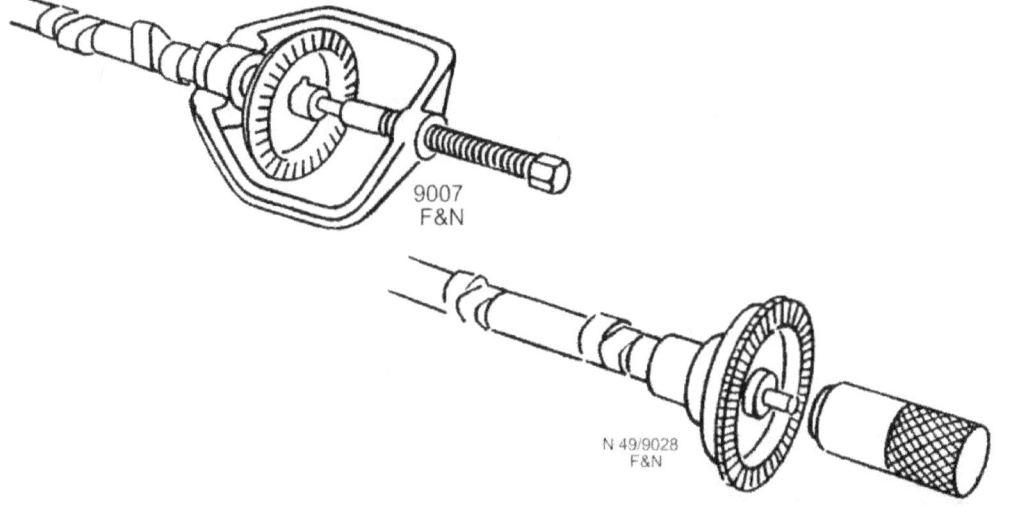

CONNECTING RODS

Engine No.:	
Year:	
Date:	Sign.:

- type of connecting rods:
  - with oil drilling _____
  - I - profile _____

oil pocket

Bearings (big end)

Inner dimension
Standard:
40.0mm + 0.036mm
− 0.025mm

		No. 1	No. 2	No. 3	No. 4
Inner dimension	A - A				
	B - B				
	ovality				

Shims

Number of shims

	No. 1	No. 2	No. 3	No. 4
marked side				
opp. side				

Bushes (small end)

reamed to 16mm
tight fit

	No. 1	No. 2	No. 3	No. 4
reamed to 16mm				
tight fit				

Assembling

	No. 1	No. 2	No. 3	No. 4
bolt threads				
nut threads				

Lubrication

	No. 1	No. 2	No. 3	No. 4
oil pockets				
oil drilling (if present)				

Final (mark with x)

| 40.0mm | 39.75mm | 39.5mm | 39.0mm |
| 38.75mm | 38.5mm | 38.25mm | 38.0mm |

Knud Jørgensen: NIMBUS - Maintenance. 2012

Connecting rods

Make a copy of the checklist 'Connecting rods'.

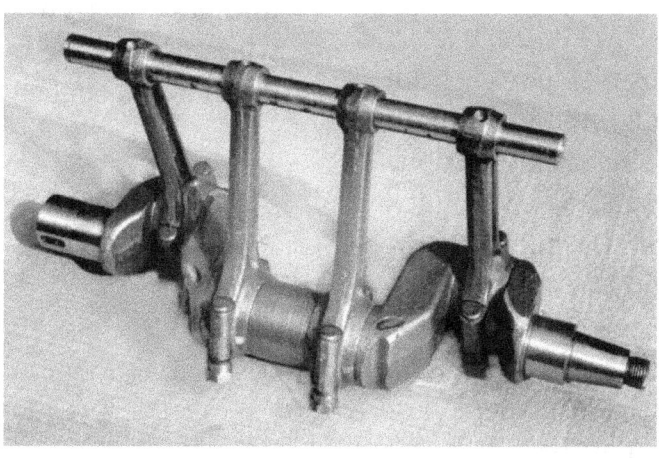

Testing the connecting rods on the crankshaft.

Factory tolerances allow the connecting rods to fall about the journal under their own weight.

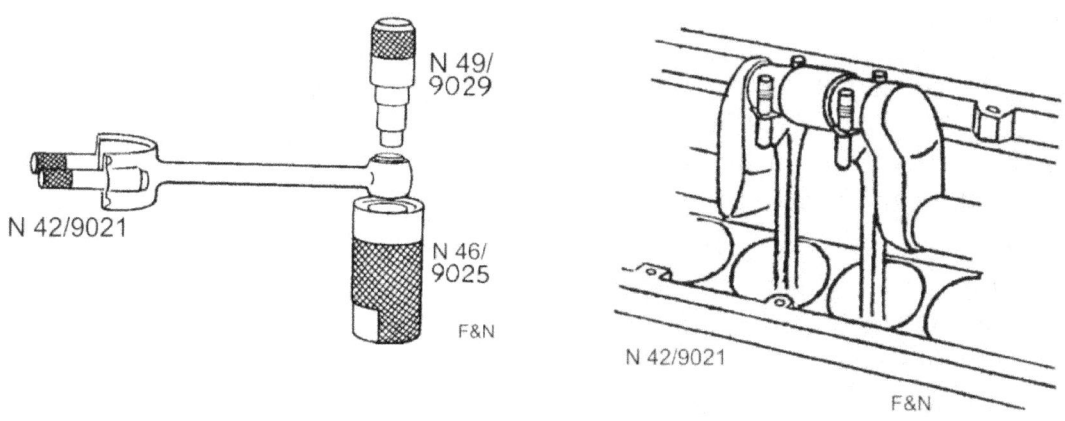

Clutch

Maintenance, repair and adjustment

The clutch transmits power from the engine through the gearbox to the rear wheel. This cannot be done efficiently if there is any clutch 'slip' under normal running conditions. Clutch slip does not usually reflect any neglect of clutch maintenance, as it is generally a result of a problem elsewhere. For example, there may be oil on the clutch plate due to any of the following conditions:

- Engine oil level being too high, or the wrong type of oil being in use

- A defective mainshaft clutch pushrod seal (7973-3)

- Worn valve guides, cylinders and rings, or a seized ring in a damaged piston, leading to crankcase pressures which the crankcase breather cannot accommodate. Oil vapour may then exit the crankcase, pass into the clutch housing, contaminate the friction material of the clutch plate, and affect clutch performance.

A temporary- but useful - solution is to rinse the clutch plate with petrol or thinners. Pour the solvent through the flywheel access hole while turning the engine over with the clutch disengaged.

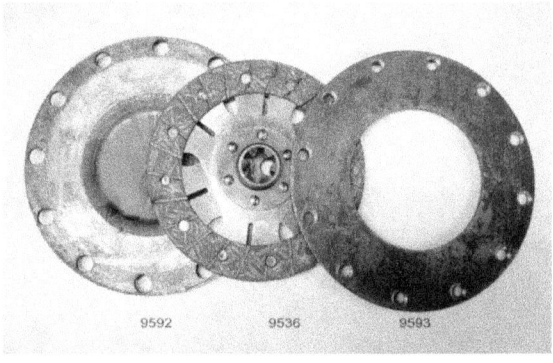

While the later clutch design, identified externally by a T-shaped release arm (9218), is especially prone to this problem, overheating and friction problems may also be experienced with the earlier clutch, which is identified by its single pivot release arm (7076)

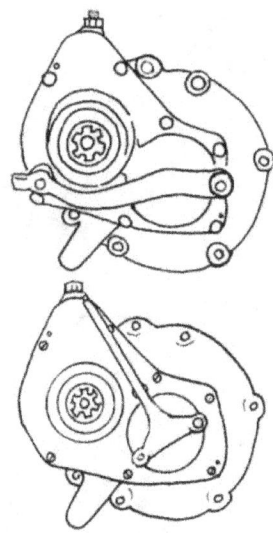

The clutch plate can deform or fracture due to over-heating and fatigue, severely affecting clutch performance. Eventually the plate may break up, rendering the clutch inoperative and possibly damaging the gearbox oil feed pipe.

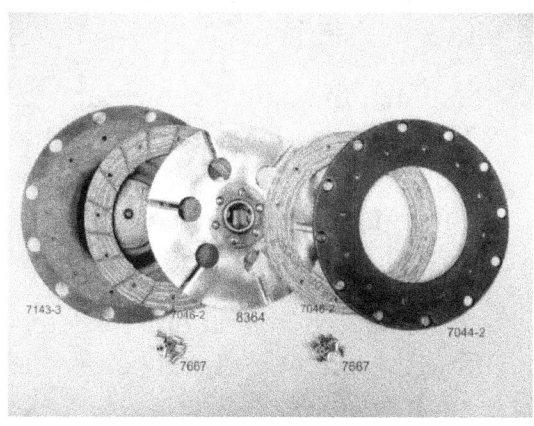

Repair

The original friction material is riveted to the clutch plate and can be renewed. Old rivets can be drilled out and new friction material fitted with new rivets. The later pattern clutch can be improved by fitting a newly manufactured plate and new clutch springs which more than meet the original specifications. New or reconditioned clutch plates are likely to have bonded friction material, removing the risk of loose rivets causing any damage.

Worn clutch hub or layshaft splines will produce clutch rattle, and may affect clutch operation, in which case replacement of the worn parts is necessary.

Adjustment

There must <u>always</u> be a clearance of 1mm to 2mm between the clutch release arm (7076 or 9218) and the external clutch pushrod (9216 or 9215), and the release arm should not contact the frame cross-member.

Combined foot and hand clutch

- Tighten the clutch pull-rod adjuster (7256), so that the release arm is correctly positioned (see above) and the clutch pedal (7281) is in its rearmost position, contacting the frame.

- Tighten the clutch cable adjuster (7208) at the handlebar just sufficiently to remove all play in the cable.

- Test clutch operation, to determine the extent of travel before the clutch releases.

- If the travel is too long, tighten the clutch release arm adjustment nut (7079) one or more half turns, making sure to leave the nut engaged with the notches in the release arm.

- If this travel is too short, (which should not be the case if clearance has been provided at the clutch pushrod), slacken the release arm adjustment nut (7079) accordingly.

Hand clutch

Early gearboxes have a clutch release arm with a single pivot (and a single adjustment nut). Later versions have a T-shaped release arm with two pivots (and two adjustment nuts). The adjustment procedures are similar.

- Tighten the clutch cable adjuster (7208) at the handlebar until the clutch cable (8413) is under tension and the clutch release lever is correctly positioned (see above).

- Test clutch operation, to determine the extent of travel before the clutch releases.

- If the travel is too long, tighten the clutch release arm adjustment nut - or nuts (7079) one or more half turns, making sure to leave the nut engaged with the notches in the release arm.

- If this travel is too short, (which should not be the case if clearance has been provided at the clutch pushrod), slacken the release arm adjustment nut - or nuts (7079) accordingly.

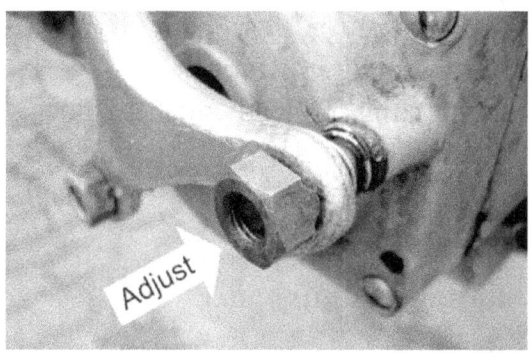

The electrical system
Battery

The battery receives and stores electrical energy. The Nimbus battery is a typical 6-volt (6V) device, assembled in a hard rubber case

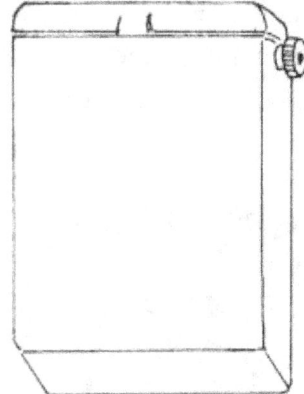

sealed with pitch. These are still for sale.

Batteries contain sulphuric acid. Even a small amount of acid will damage skin, clothing, paintwork and polished aluminium. During charging, a mixture of hydrogen and oxygen gases may be given off by the battery. This mixture is explosive. Do not use a flame or any device which could ignite this mixture when checking the acid level in the cells.

Sealed batteries in black or grey nylon cases are now available. These are less expensive than the older type of battery and are maintenance-free. They are of excellent quality. A black 6volt 16Ah version, with lid, is usable as is. And a cheaper, grey version (6V 14Ah) can be concealed inside a modified older type battery case.

Removing the battery

- Disconnect the leads at the battery terminals (earth lead first) to prevent any short circuit during removal.

- Remove the 2 knurled nuts (7526) and the retaining bracket (7553) and U-bolt (7550).

- Lift the battery out between the saddle and the pillion seat. With a spring-mounted saddle there may be little room. The springs may have to be detached from the frame so that the saddle can be lifted a little. Avoid tilting the battery and spilling acid.

Fitting the battery

Follow the removal instructions in reverse order. Make sure *before* connecting the leads, that the positive (+) terminal will be connected to 'B' and the negative (–) terminal to the frame.

Battery maintenance

Batteries have a limited life. The plates will sulphate in the course of time and eventually there may be a short circuit in one or more cells. The better a battery is maintained, the longer will be its life. When just one of the cells has failed, the battery must be immediately replaced. Running with a defective battery will overload the dynamo.

The acid level in each cell has to be checked regularly. It should be a few millimetres over the plates, and if not it must be topped-up with demineralised water. Do not top-up with acid, because it is only water which has been lost and must be replaced. If frequent replenishment is required under normal operating conditions, this indicates that the regulator may need to be adjusted. The charging current may be too high, leading to 'gassing' (the giving off of hydrogen and oxygen through electrolysis).

A battery which is not in use - for example over winter - should be disconnected and stored in a cool and dry place. It should be charged regularly, but not overcharged.

Dynamo

Removing the dynamo
See dismantling the engine.

Dismantling the dynamo

- Remove the 4 screws (5400) retaining the brush holders (7870).

- Remove the <u>left</u> brush holder (8097) with its external earth loop (bare copper wire) and the left brush.

- Remove the 'D' brush from the <u>right</u> brush holder. *Note:* Do *not* attempt to remove this right brush holder until the dynamo has been further dismantled.

- Remove the upper and lower split pins (3866 and 4518) from the armature.

- Remove the upper castellated nut (7239). Use special fixture N15/9002 to hold the dynamo or clamp the lower castellated nut in a vice. *Note:* Do *not* clamp the shaft end directly. If the special castellated socket N 16/9003 is not available use long-nosed pliers or a punch.

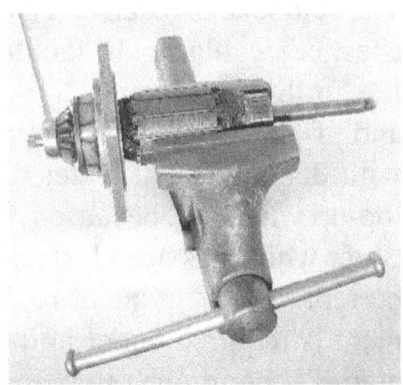

- Remove the upper bevel gear and oil slinger (8056 or 8056-2) from the armature. A puller may be needed.

- Remove the key (7311) from the armature, using flat-nose pliers.

- Take the dynamo out of special fixture N15/9002 or out of the vice.

- Remove the 2 nuts (7258) that retain the dynamo lower end bracket (7880)

- Remove the dynamo end bracket complete with castellated nut, bevel gear and armature from the upper end bracket and dynamo body. *Take care:* Pull out the small dowel (7914) which will have been retained in either the dynamo body or the end bracket.

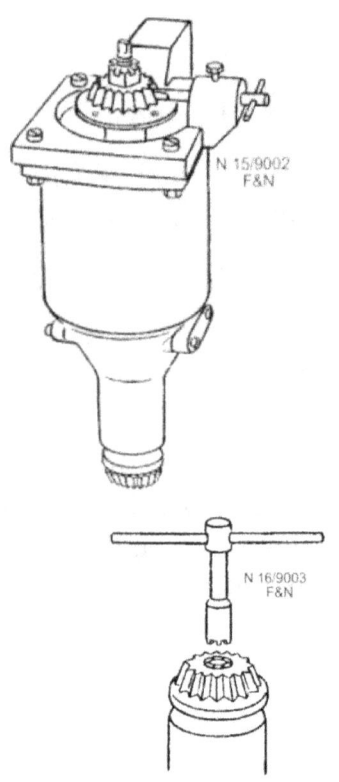

- Use special fixture N15/9002 to hold the armature, lower end bracket, castellated nut and bevel gear. If the fixture is not available, clamp the armature in a vice.

- Remove the lower castellated nut. (7238).

- Remove the bevel gear (7147) from the armature.

- Remove the key (7311) from the armature using flat-nose pliers

- Separate the lower end bracket (7880) and the armature.

- Remove the 4 screws (7500) securing the retainer plate (7474) to the end bracket and remove the ball bearing (7316/SKF 6301)
.
- Disconnect the field coil leads from inside the D-F brush holder. *Take care:* Check first that the connectors are marked 'D' and 'F'. If not, mark them immediately.

- Remove the two ball bearings (7315/SKF 6201) and the oil slinger (7884) from the upper end bracket, or dynamo 'neck' (7879).

- Remove the 2 long studs (7885) which connect the upper and lower dynamo end brackets.

- Take out the 2 insulators (8459) from between the field coils (7913) in the dynamo body (7917).
Dismantling should continue only if the parts concerned require renewal or repair.

- Remove the 4 screws (7918) that secure the field magnets (7882) to the dynamo body. Use an impact screwdriver.

- Remove the 2 field coils (7913) and field magnets.

- Separate the field magnets from the field coils.

Assembling the dynamo

Follow the dismantling procedure in reverse.

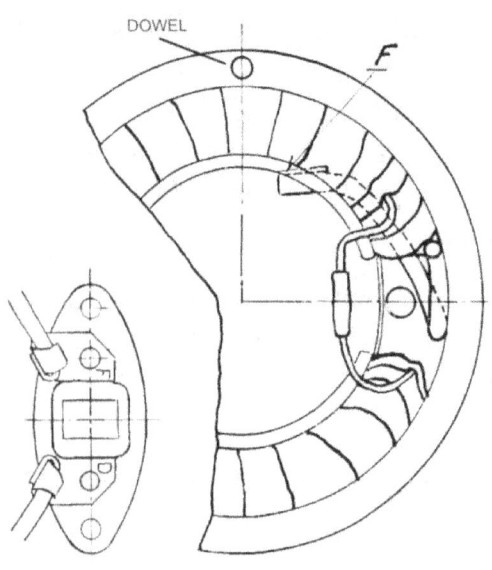
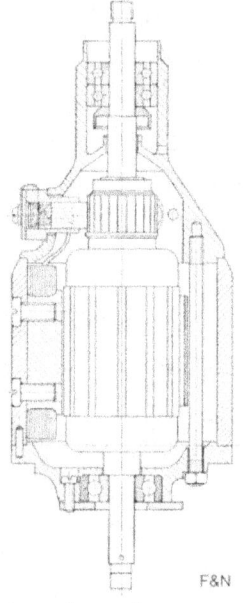

Dynamo maintenance, repair and testing

The dynamo will not require a lot of maintenance provided that the battery, regulator and other components of the electrical system are in good order. Pay attention to the electrical demand on the dynamo. By this is meant the power consumption of the head light, taillight, side car light and any other electrical components. Make sure that all earth connections are sound and solid and that all leads are of the right specification (1.5 mm²).

Dynamo brushes do wear, but in a well-functioning circuit, only very little. If the brushes wear noticeably and require regular renewal, this is a sign that something is wrong. This is not necessarily a dynamo fault, but may be the result of a problem elsewhere in the electrical system.

The dynamo brushes can be changed with the dynamo in place. First, remove the oil feed and return pipe from the camshaft housing. Remove the left brush holder with the earth connection (bare copper wire loop) and dynamo brush. Remove the 'D' brush at the other side together with its lead. This allows access to the commutator, which can be cleaned with a petrol soaked cloth while turning over the engine with the kick-starter.

If the dynamo brushes do not move freely in their holders despite spring pressure, this is usually caused simply by a build-up of oil and carbon dust, but traces of solder splashes are are a sign that soldered connections on the armature have melted. This condition requires total dismantling and a complete dynamo overhaul.

Repair of the dynamo on the road (emergency repair)

First, remove the oil feed and return pipe from the camshaft housing. Remove the left brush holder with the earth connection (bare copper wire loop) and dynamo brush. Remove the 'D' brush at the other side together with its lead. This allows access to the commutator, which can be cleaned with a petrol soaked cloth while turning over the engine with the kick-starter.

Gently burnish the commutator with a piece of fine abrasive paper or cloth. Then with a sharp tool, undercut the insulation between the segments to a depth of ½mm. Clean both brushes thoroughly in petrol and rotate each in its holder when fitting.

Once at home, the dynamo should be removed, taken apart and repaired. But remember that dynamo problems may be the result of faults elsewhere in the electrical system.

Repair of the dynamo

A dynamo armature in very poor condition must be totally overhauled at a specialized workshop. This involves rewinding the armature and replacing the commutator.

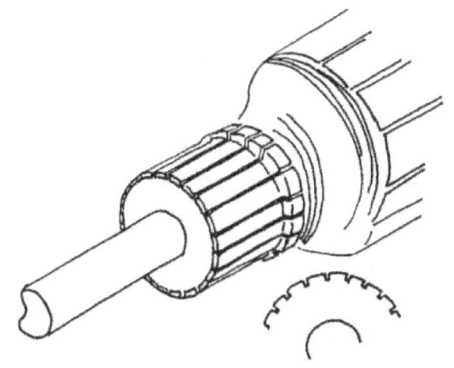

However, it may be possible to home-repair melted connections between the armature and commutator as long as the commutator has not worn to a diameter less than 32mm. If a test (see applicable section) proves after such repair that the armature windings are in order, the commutator can be machined on a lathe and the insulation between segments carefully undercut with a hacksaw blade to a depth of ½ mm.

Note: It is important to check that all grit has been removed. Even with newly overhauled armatures it may occur that the gaps between the segments are not totally free of foreign matter. Any grit remaining may dislodge the brush or cause arcing which can damage the commutator.

Field coil windings can short circuit or burn out interally. But if there is no continuity through the field coils a checks should be made of the short lead between the coils and of the leads to the 'D' and 'F' terminals.

The keys and the keyways in the armature can wear - especially those at the upper bevel gear. If so, the keyway in the bevel gear will certainly also be worn. It is possible to renew keys and bevel gears but if a keyway in the armature is damaged, this must be rectified or the entire armature has to be renewed. Any clearance, allowing any movement of a bevel gear around the shaft, can disturb the ignition timing.

The armature drive spade for the oil pump may also wear. In the worst case, the drive spade may damage the drive slot of the oil pump gear leading to oil pump and engine failure. The drive spade can be built up by welding and machined back to size.

Take care: The brush holders are retained by 4mm x 18mm screws - if they are any longer than this, they may short to the commutator.

Testing a dismantled dynamo

Check for field coil continuity using a meter or battery and test lamp. Check the electro-magnetic field by flashing the 'D' and 'F' leads to a battery *for a short period of time.* A screwdriver shaft should be attracted by either field magnet.

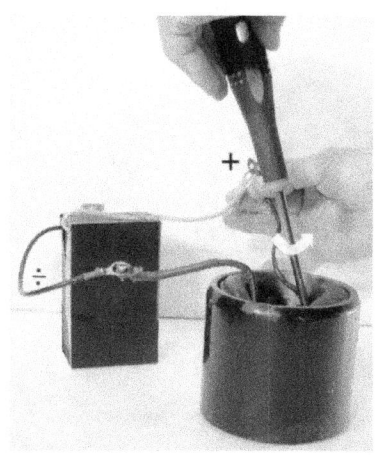

Check the armature with a battery and a test lamp. There must be continuity between every single segment of the commutator and all other segments. *Note:* This test does not detect a fault in the armature windings.

Testing an installed dynamo

The dynamo can be tested when installed. Equipment needed is a voltmeter, an ammeter and a short cable with a crocodile clip at each end. Remove the 'D' and 'F' leads from the right hand brush holder leaving the 'D' brush in place.

Measuring residual voltage:

The residual voltage is the voltage generated solely by the residual (permanent) magnetism present in the field magnets when electrical power is not supplied to the field coils. Connect the positive (+) lead of the voltmeter to the 'D' brush terminal and negative (-) lead to an earth point on the frame. Start the engine and note the highest recorded voltage. It should be about 1.5V. If there is no reading or, if the voltmeter reads below zero, the dynamo has to be repolarised (See the applicable section). If there is still no reading after this process the armature is faulty.

Measuring maximum voltage:

Connect the positive (+) lead of the voltmeter to the 'D' brush terminal and negative (-) lead to an earth point on the frame. Connect the 'F' terminal also to an earth point the frame, using a short lead. Start the engine and slowly increase the engine speed until the highest voltage has been reached. It should be 24–28V. Anything less indicates a fault in the armature and/or the field coils.

Measuring the maximum charging current:

Connect the 'D' terminal to the positive (+) battery terminal, and the 'F' terminal to an earth point on the frame by means of a short lead. Connect the ammeter across the 'D' terminal and the battery positive. The dynamo should produce about 10 amps. Any reading less than this indicates a fault in the armature and/or the field coils.

Take care: The output of the dynamo is at its maximum during this test. To avoid damage to the armature do not run the test any longer than is absolutely necessary.

Testing the dynamo when not installed

An assembled dynamo which is ready for use can be tested before installation.
- Connect the negative (-) terminal of the battery to a bare metal part of the dynamo body. Connect the battery positive (+) terminal to the 'F' terminal of the brush holder.

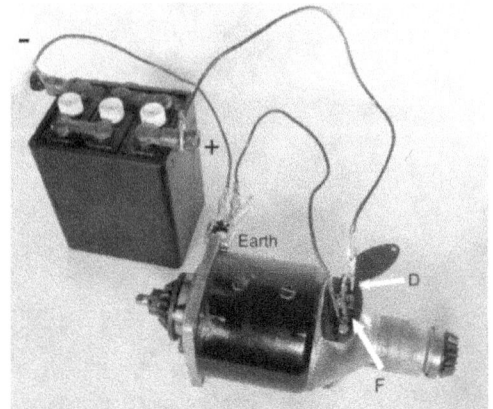

If the armature rotates *slowly* <u>against</u> its normal direction of rotation - a slight start by hand may be needed - then the field coils are in good order. (The normal direction of rotation is indicated by an arrow on the edge of the lower end bracket).

- Connect the negative (-) terminal of the battery to a bare metal part of the dynamo body to the 'F' terminal of the brush holder. Connect the battery positive (+) terminal to the 'D' terminal of the brush holder. If the armature now runs full speed in the <u>normal</u> direction of rotation - a slight start by hand may be needed - then the armature is working correctly.

(The normal direction of rotation is indicated by an arrow on the edge of the lower end bracket).

Polarising the dynamo

The dynamo can lose its polarisation. If the battery and the wiring harness are in good order and the battery charge indicator light does not go out as engine speed increases, depolarisation could be the cause.

With the engine running, briefly connect the battery positive (+) teminal and the 'D' terminal of the dynamo. The charge indicator lamp should go out and stay off when the connection is removed. If that is not the case, the problem is not depolarisation, but a faulty armature and/or field coil.

To repolarise the field magnets, connect the 'B' and 'D' terminals of the regulator (A/S Fisker & Nielsen or Bosch – see below).

Note: With an electronic regulator, repolarisation requires by-passing the regulator.

Voltage Regulator

A/S Fisker & Nielsen used two different makes of voltage regulator, first their own (7939 and 7939-2) and later a Bosch unit (10671). (The oil pressure cut-out - used only in 1934 - is not described here.)

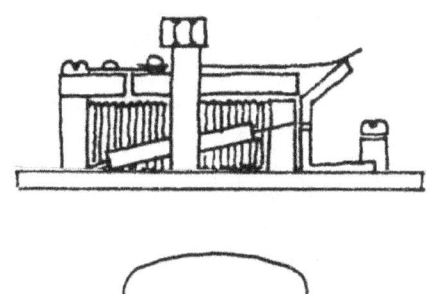

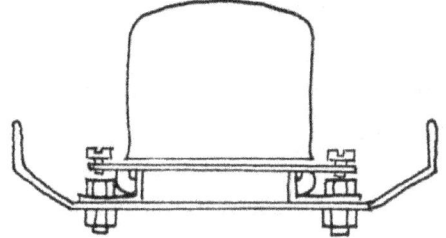

Removing the voltage regulator

Before working on the regulator, disconnect the earth lead from the battery to the frame.

- Disconnect all leads connected to the regulator. If the leads do not have different colours or do not have numbered terminals, attach a clear label to each.

- Remove the two bolts (7794) securing the regulator bracket (7937 or 10669) to the frame.

Fitting the voltage regulator

Follow the removal instructions in reverse order.

Take care: Connect lead 'D' to terminal 'D' (the lead may be blue), and lead 'F' to terminal 'F' (the lead may be yellow). For a Bosch regulator connect lead 'D' to terminal 'D' and lead 'F' to terminal 'DF'. If the leads are reversed, the regulator will be damaged immediately it is connected.

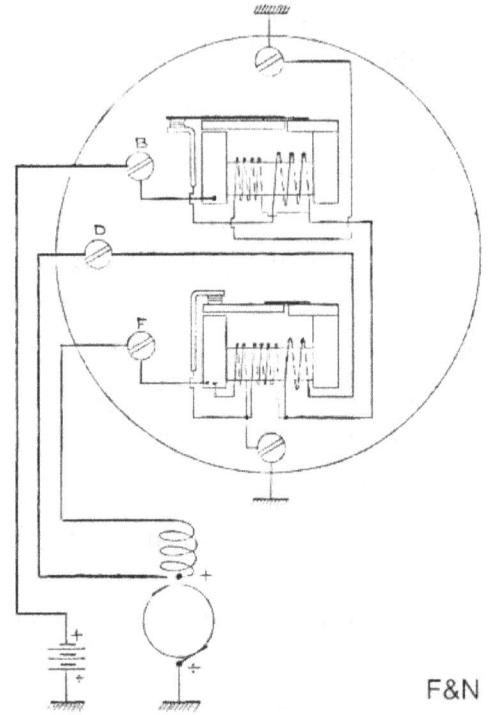

F&N

The bracket must be well earthed to the frame. Star or serrated lock washers are useful. If the frame has been newly painted, scrape off some paint from around the bolt holes: the earth connection must be to bare metal.

Repair and maintenance

The cover of an F&N regulator is attached by one or two screws (5532) each with a cup washer (8118) which was originally sealed and stamped F&N. (A broken seal rendered the warranty invalid). Do not tighten or slacken anything in the regulator. Adjustment instructions can be found below.

The regulator must be kept clean, especially the connections. It may be necessary to clean the contacts from time to time with a piece of fine emery. Adjustment can be carried out in accordance with the instructions of the factory, or can be left to a specialist. Repair of an electro-mechanical regulator that is 60 or 80 years old, is normally not worthwhile. Replacement is recommended, using an electronic regulator.

The factory issued instructions in 1935 on how to adjust the original regulator. These are reprinted below.

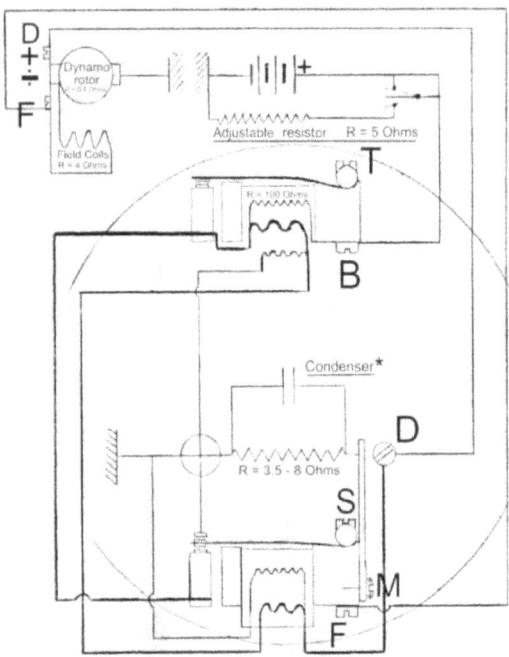

* Condenser Capacity (in 1934 shown as 20.000 cm) = 0.02 mF

Adjusting the regulator

The regulator unit includes two separate relays, both of which are adjustable. One is the cut-out, which protects against reverse current flow. The other is the voltage regulator, which controls the rate of charge to the battery.

The regulator terminals must be connected as follows:
B to the battery positive terminal
D to dynamo terminal **D**
F to dynamo terminal **F**

The earth connection is by one of the regulator bracket fixing bolts.

Before any adjustment, disconnect the battery and replace it by a 5 Ω test resistor. Check all adjustments again with the battery in place.

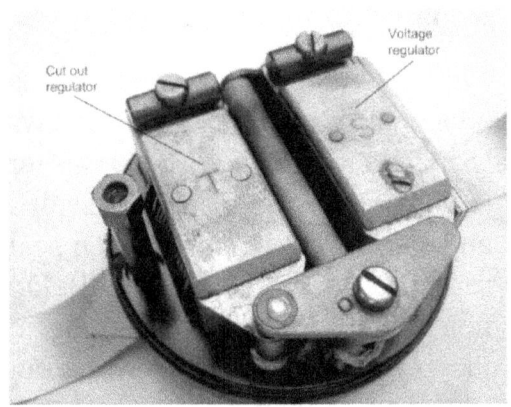

Adjusting the cut-out:

The cut-out disconnects the dynamo from the battery when the dynamo output is lower than the battery voltage.

Set adjusting screw **T** such that the cut-out contacts close when the dynamo output measures:
- 6.0 volts (initial setting, from cold)
- 7.0 volts (warm setting)

Adjusting the voltage regulator:

Correct setting of the voltage regulator ensures that the battery is kept fully charged but not overcharged.

Set adjusting screw **M** temporarily such that it protrudes about 2mm. With the motor idle, turn adjuster **S** clockwise, intil the dynamo voltage (with the voltage regulator working on its upper contact) is:
- 6.2 volts (initial setting, from cold)
- 7.2 volts (warm setting)

As engine speed is increased, the voltage regulator will switch to its lower contact.
Adjuster **M** must be set such that the voltage is then 0.3 volts higher.

Finally; **S** is adjusted second time, to ensure that the voltage is:
- 6.2 volts cold and 7.2 volts warm with the upper contacts closed; and
- 6.5 volts cold and 7.5 volts warm with the lower contacts closed.

Ignition coil and distributor
Maintenance

Ignition coil has to be handled with care, especially while removing or fitting it. Bakelite becomes less stable over time. Keep the ignition coil as clean and dry as possible. A combination of contamination, moisture and the high voltage of the ignition coil can result in surface arcing tracks which will reduce the insulating properties of the coil housing. The distributor contacts have to be checked at regular intervals and any traces of arcing at the points or on the rotor or the rotor contacts have to be removed. In addition, the high-tension lead connections at the distributor cap have to be regularly checked, cleaned and sprayed with a moisture repellant. The spring clips of the plug terminals can lose their effectiveness. Renew them when the distributor contacts are replaced. Do not try to take apart the ignition coil assembly. Both the coil and the screw-attached cover are sealed with pitch.

Any deterioration of the insulation in the coil will reduce the sparking voltage, and this effect is noticeably increased when the coil is hot. If the ignition coil fails periodically, perhaps after a few kilometers, and then seems to recover, this indicates a winding or insulation fault. As the coil heats up, the resistance of the copper windings increases and this reduces the high-tension voltage (the sparking voltage). Misfiring and ignition failure are the result. The coil may function again when its internal temperature falls, but ignition failures are likely to become more frequent and the coil should be replaced.

Testing

Ignition coil performance can be tested by dedicated coil checking equipment which subjects the coil to demands similar to those which apply in use. The ignition coil should be able to generate sparks across a 10 mm air gap for at least 10 minutes. A faulty coil may work perfectly well for several minutes, but as it warms up the sparking voltage may fall, resulting in a weaker spark and finally in no spark at all.

Repair

Small chips or cracks in the ignition coil housing or cover can be repaired with two-component resin. Traces of any arcing tracks have to be cleaned thoroughly with solvent and filled with resin. The four distributor high-tension rotor contacts and the four high-tension lead sockets have to be renewed together. Drill out the rivets of the contacts and use a thin punch to free the sockets from them. Fit new sockets and contacts, and rivet all in place. Suitable screws, lock washers and nuts can be used in place of rivets. A defective ignition coil can be extracted from the coil housing, but this has to be left to a specialized workshop, where a new coil can be wound and fitted. New ignition coil assemblies are still produced for sale.

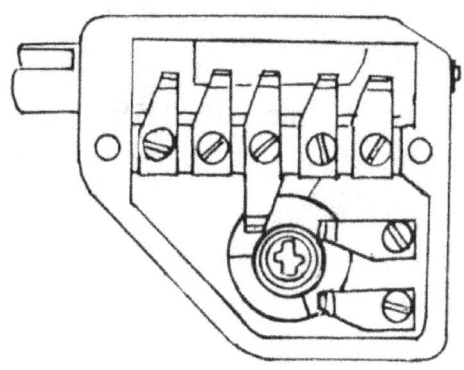

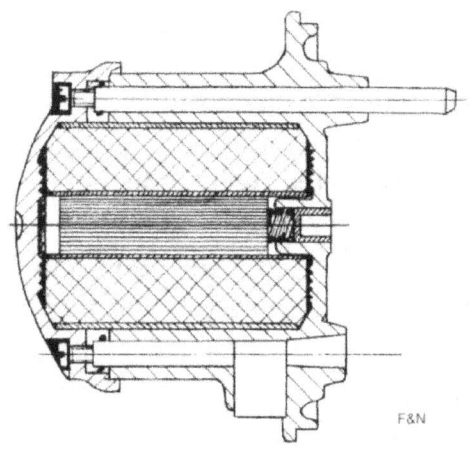

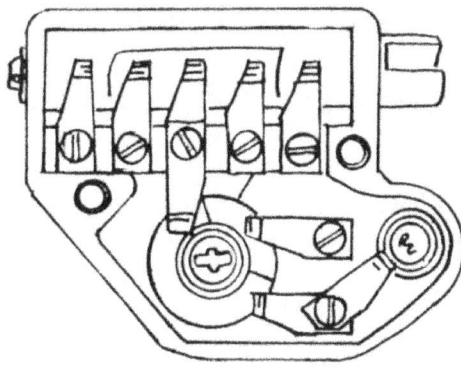

Combination switch

There are two versions of the combined ignition and lighting switch for the Nimbus-C, due to changes in the handlebar pressing. The early switch (7345) fits the flat handle bar used with the 'early' internal slider forks, and the later version (8880) was fitted to machines with the revised handlebar and 'later' external slider forks.

Dismantling the combination switch

- Remove the switch (see dismantling the handle bars)

- Remove all contact screws (5532 or 5508) and all wiper contacts (7489 or 8901).

- Remove the ignition switch rotor (7336) from the switch housing. - - The conical spring (7780) in the rotor can be prised out with a small screwdriver

- Remove the 6V, 1.5W charge warning bulb - Philips 6876. (Later switch only).

- Remove the split pin (3867) and washer (5943) from the end of the switch drum in the switch housing.

- Pull the switch drum (7367 or 8871) out of the housing.

Assembling the combination switch

Follow the dismantling instructions in reverse order.

Take care: The conical spring has to be fitted in the ignition rotor with the smaller diameter uppermost. (That is, <u>not as illustrated</u> in the spare parts list). Note that the wiper contact (8901) of the charge light bulb is different: it has a cut-off corner.

Repairing the combination switch

Worn wiper contacts:
- If the wiper contacts working areas are worn through, or are soon expected to be, then the contacts must be renewed.

Worn switch drum:
- If the wiper contacts have caused grooving of the switch drum, it can be *carefully* dressed. This can be done by clamping the drum in a drilling machine and applying fine emery.

Worn or broken ignition switch rotor: renew the rotor.

- Cracks in the switch housing: The housing is a bakelite moulding and can be repaired with two-component resin. It is possible to replace failing or damaged threaded bushes, also by using two-component resin.

The threaded bushes in the early switches are are moulded in the bakelite, but in later switches they are pressed into the material and can therefore also be pressed out when replacement is necessary.

Ammeter and charge warning light

The connections for the ammeter (7665) are shown in the wiring diagrams for machines numbered 1301 to 1550 and 1551 to 2400. The ammeter is fitted in the left of the handlebar and shares the speedometer illumination. It needs no maintenance and can be repaired when necessary.

The connections of the charge warning lamp (8290) are shown in the wiring diagram for machines numbered 2401 to 7500. The upper part of the lamp is clipped into the circular name and number plate (8287 + 8288) on the handlebar.

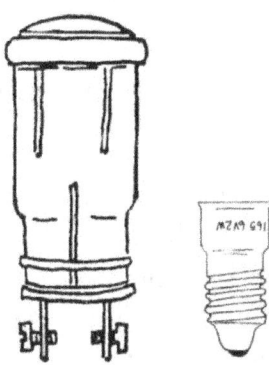

The charge warning light shown in the wiring diagram for machines numbered 7501 to 14015 is also fitted in the handlebar (see combination switch).

Horn and horn button

The Nimbus-C had various horns fitted, but all have an electro-magnetic coil which is activated when the horn button is pressed, causing an armature to be attracted to it. The armature is fitted on a diaphragm and as soon as it is attracted, power to the coil is cut off. The armature springs back and closes the contact again. Rapid oscillation of the diaphragm which causes vibrations in air which are perceived as sound.

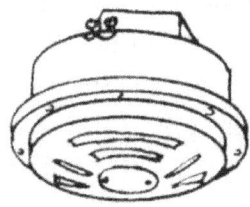

Removing the horn

- Disconnect the leads to the horn.

- On 'early' internal slider fork machines, remove the two screws (7265) which secure the horn bracket (7381-Ri), to the front of the handlebar.

- On 'later' external slider fork machines remove the two screws (7265) which attach the horn bracket (9585 or 7381-H) to the fork yoke.

Fitting the horn

Follow the removal instructions in reverse order.

Dismantling the horn

The Riemann horn is rubber mounted and is retained in its mounting by a circular clamp (7381-Ri). Loosen the clamp and remove the rubber ring. Most horns can be dismantled by removing screws around the rim. This allows removal of the front cover and diaphragm, giving access to the contacts and condenser. Do not dismantle the horn further. Normally, it is only the contacts or the condenser that need attention.

Maintenance and repair

If the horn fails, the cause may be a fault in the horn button, or in the wiring from the combination switch to the horn or from the horn button to the horn. Check with a test lamp for power to the horn. If the horn is powered but does not work, it is likely that the contact points in the horn are at fault. Dismantle the horn and clean the points.

If the points are burnt, it is probably due to a missing or failed condenser.

If the horn works but the sound is poor, check the voltage to the horn. If it is below 6 volts the battery should be charged. If the voltage is sufficient then adjust the points by means of the adjustment screw. (This is the *off-centre* screw on the back plate). Turn the screw ¼ turn at a time, in or out, and test the horn.

Removal of the horn button

Removal of the push button (7385) from the early handlebar: Remove the top of the housing by unscrewing it or by releasing the spring clip, take out the push button and spring and remove the 2 screws (5249) in the base of the housing.

Removal of the horn button (9324) from the later handlebar: remove the 2 screws (5509) of the horn push button.

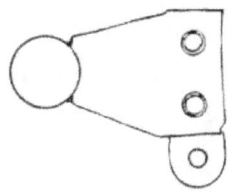
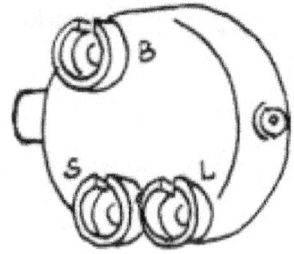

Fitting the horn button

Fitting of the horn button (7385) on the early handlebar: Fix the push button housing to the handlebar by means of the 2 screws (5249) in the base, connect the lead from the horn, and fit the spring, push button and the top of the housing, which is either screwed into place or retained by a spring clip.

On the later handlebar, attach the horn button (9324) with 2 screws (5509).

Brake light switch

The brake light switch (7764) is fitted in an exposed position and there is a risk of moisture leading to corrosion at the terminals and contacts.

Removing the brake light switch

- Disconnect the battery earth lead.

- Remove the pull spring (7769) of the brake light switch from the brake pedal clevis pin (7774).

- Mark the leads with 'B', 'S' and 'L' and remove them.

- Remove the switch retaining nut (7258) and remove the switch.

Fitting the brake light switch

Follow the removal instructions in reverse order.

Dismantling the brake light switch

- Release the hook of the pull spring from the rod (7767 or 7767-2).

- Pull the rod from the switch housing.

- Carefully release the switch return spring (7773). This releases the insulation washer (7570), contact spring (7770) and insulating bush (7768).

Assembling the brake light switch

Follow the removal instructions in reverse order. *Take care:* the spring contact must be correctly oriented and fitted onto the shoulder of the insulating bush. After assembly, use a test lamp to check that when the switch is operated no electrical connection is made other than between the 'B' and 'L' terminals. This is essential.

Maintenance

Regular use of a moisture repellent spray is necessary on this switch. Removal, disassembly, and cleaning of the brass terminals has to be done annually. If the spring contacts are excessively bent (see drawing) correct this. *Take care:* Spring bronze is subject to fatigue.

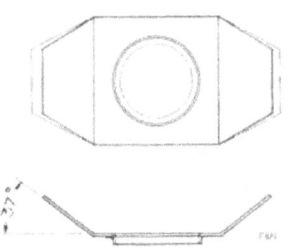

Lighting equipment

Headlight

Headlights of various makes and types have been fitted. Partial dismantling for maintenance purposes can be done without removing the headlight.

Dismantling and assembling the headlight

- Disconnect the battery earth lead.

- Remove the rim and reflector with retention spring and bulb holder.

- Detach the retention spring (if present) and remove the bulb holder.

- Remove the headlight bulb and parking light bulb (if present).

- The speedometer can be removed once the drive cable has been disconnected

- To assemble, follow the dismantling instructions in reverse order.

Removing and fitting the headlight

- Disconnect the battery earth lead.

- Disconnect the leads to the headlight at the handlebar combination switch

Remove the headlight fixings at the fork-mounted headlight brackets or on the lower fork yoke and remove the headlight.

- To fit the headlight, follow removal instructions in reverse order.

Note: it is good practice to fit an earth lead directly from the headlight bulb-holder to an earthing point on the motorcycle frame. If a wiring harness is being made, it is best to include an earth lead directly connecting the headlight bulb-holder to the negative (-) battery terminal. An earth to the fork or to the fork yoke is not reliable.

Tail light

Partial dismantling for maintenance purposes can be done on the fitted tail light.

Dismantling and assembling the tail light

- Disconnect the battery earth lead.

- Remove the spring ring (7571 or 10991) and the tail light lens (7572 or 10983).

- Take out the aluminium (8180) and the celluloid segment (7573) from the tail light housing.

- Remove the tail light and brake light bulbs.

Further dismantling requires removal of the tail light body from the mudguard.

- Remove the two screws (5350 or 5253) securing the terminal plate (7695).

- Remove the spring contacts (7568) from the terminal plate (7695).

Removing and fitting of the tail light

- Disconnect the battery earth lead.

- Slacken the two rear wheel spindle nuts and hinge up the rear mudguard. Keep it in this position by a strap or cord to the steering damper knob.

- Disconnect the tail and brake light leads from brake light switch terminals 'S' and 'L'.

- Detach the tail and brake light lead from the single retaining clip (4571) inside the frame and the three similar clips inside the rear mudguard.

- Remove the 2 screws (7265 or 7168) securing the tail light body and rubber mounting pad (8182 or 10725) to the mudguard (8173)

Sidecar lights

Sidecar lights of various makes and types have been fitted. Partial dismantling for maintenance purposes can be done without removing the light.

Dismantling and reassembling sidecar lights

- Disconnect the battery earth lead.

- Remove the lens or the rim and lens if necessary.

- Remove the light bulb. If the bulb is difficult to remove because of corrosion, clean the area with parrafin (kerosene). Remove the complete light if necessary. This usually means removing the sidecar wheel.

- Assemble in the reverse order. Make sure, especially when newly sprayed parts are fitted, that good earth connections exist between the motorcycle and sidecar frames and between the sidecar frame and mudguard. Fitting a direct earth lead connection will result in a more reliable connection.

Wiring

The original wiring in the Nimbus-C electrical system was rubber-insulated and linen-wrapped. These materials require meticulous maintenance to avoid damage from oil, petrol and heat, and suffer from ageing. Later wiring harnesses use PVC insulation (which is much superior) and may be colour coded..

Removing and refitting the wiring harness

- Always disconnect the earth lead from the battery terminal first when working on the electrical system. A short circuit with a fully charged battery can burn out the wiring harness in a matter of seconds.

- Remove the secondary (lighting) harness first. This involves connections to the battery, regulator, brake light switch and tail light.

- Then remove the main (ignition and charging) harness. Refer to the applicable wiring diagram.

Fit the harness in the reverse order.

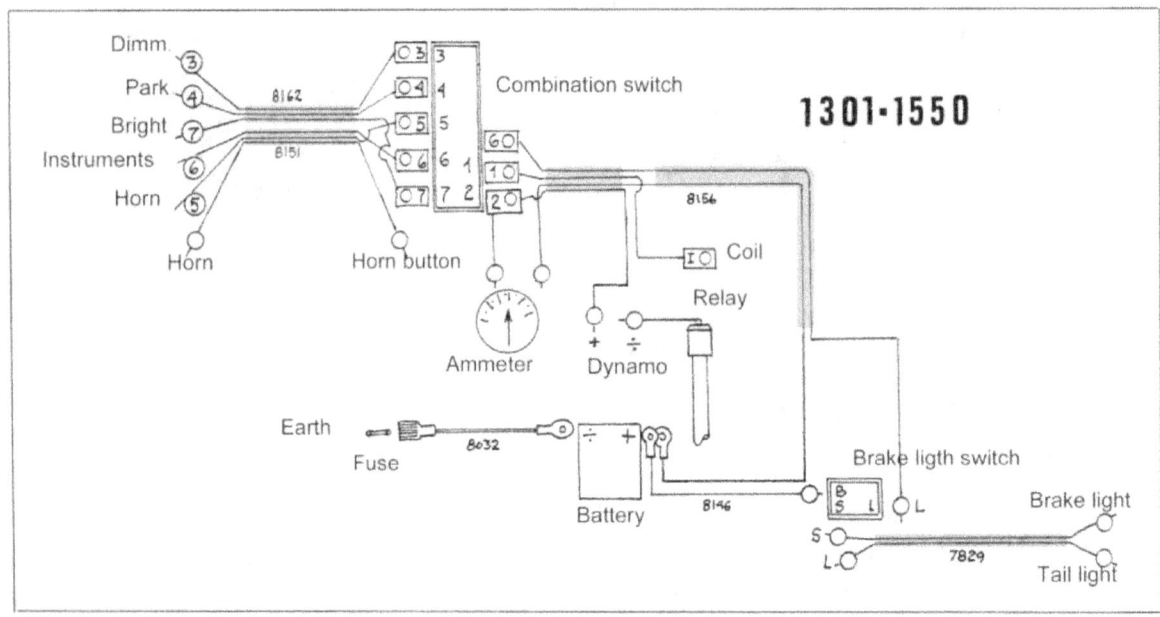

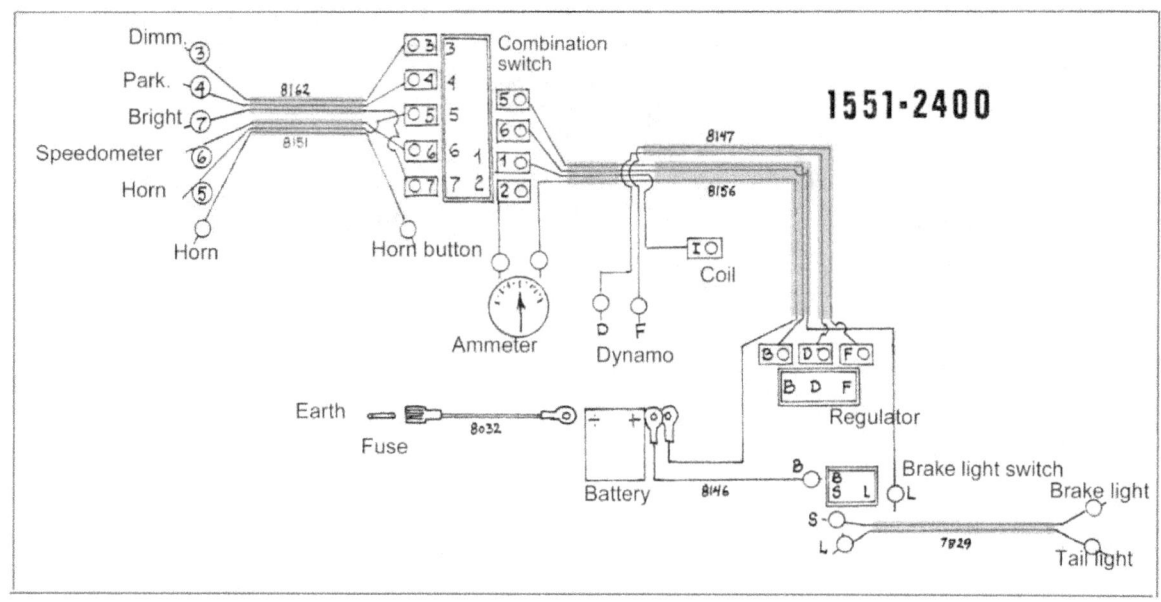
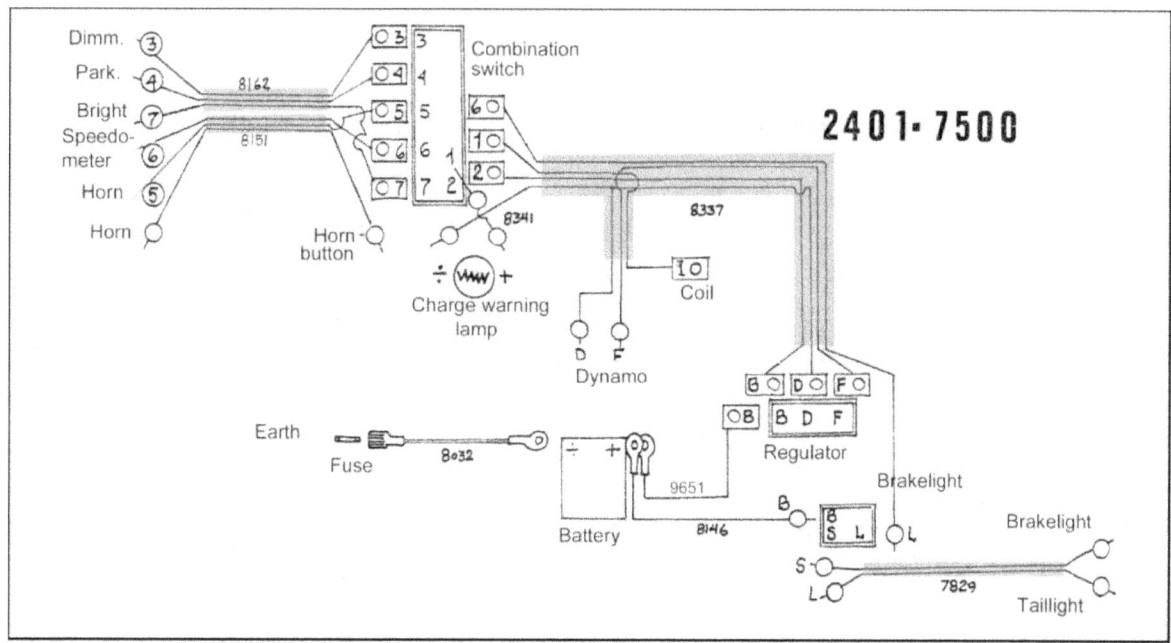

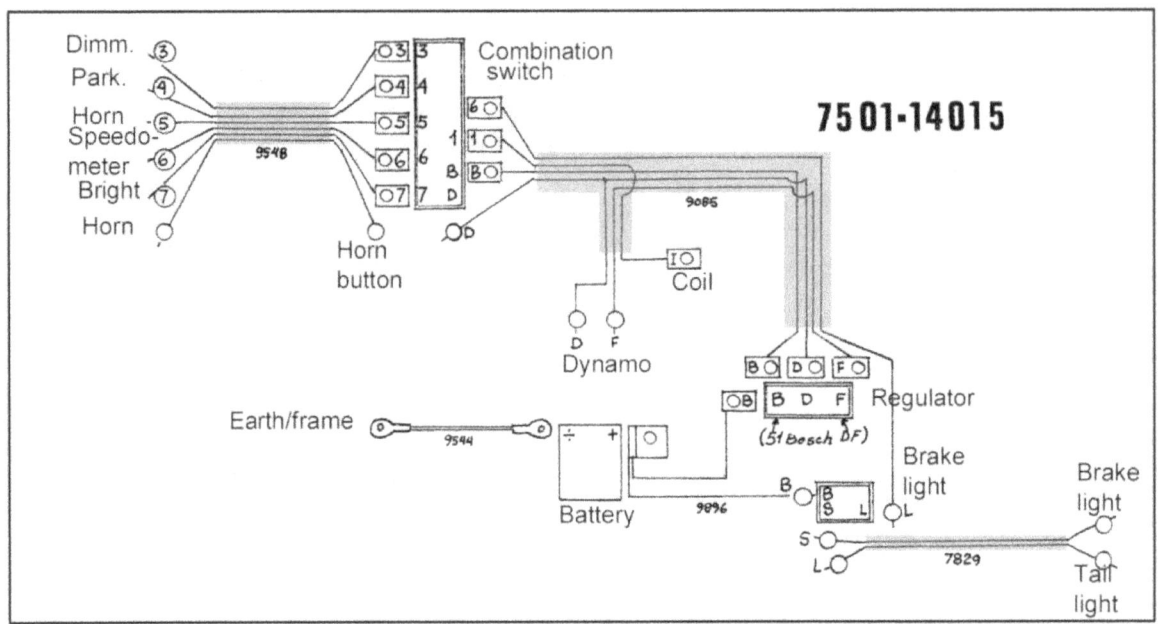

Fitting a colour coded wiring harness

- Connect the main (ignition and charging) harness (9085) to the combination switch (8880) and brake light switch (7765) as follows:

    - Blue to 'D' (dynamo)
    - Yellow to 6 (lights)
    - Red to 'B' (battery)
    - Green to 'I' (ignition)
    - Black to 6 (lights) and to the brake light switch 'L'.

- Connect the main harness to the dynamo as follows:

    - Blue to 'D',
    - Yellow to 'F'.

- Connect the main harness to the ignition coil, Green to 'I' (on the coil).

- Connect the main harness to the regulator (7939, 7939-2 or 10671) as follows:

    - Blue to 'D' or 'D+'
    - Yellow to 'F' or 'DF'
    - Red to 'B' or '51'.

- Connect the secondary (lighting) harness from the headlight/speedometer to the combination switch as follows:

    - Blue to '7' (high beam)
    - Red to '3' (low beam)
    - Yellow to '6' (lights)
    - Green to '4' (parking)
    - Black to '5' (horn) and to the horn button.

- Connect the lead (9896) between the regulator and battery.

- Red/black from the regulator to the battery positive (+) and from the battery positive (+) to brake light switch terminal 'B'.

- Connect the twin core lead (7829) from the brake light switch 'L' terminal to the tail light bulb contact, and from the brake light switch 'S' terminal to the brake light bulb contact.

- Fit lead (9544) from the battery negative (-) terminal to the frame.

Repairing the wiring harness

All connectors in the wiring harness should be soldered. For crimped connectors, special-purpose crimping pliers must be used. Connectors crimped with general purpose pliers can fail with use.

Use the correct gauge automotive cable (1.5mm² / AWG 14) for all leads. Telephone and loudspeaker wires for example, have inadequate carrying capacity for current levels in a motorcycle electrical system. Ensure that terminals and leads are coded with numbers, characters or colours consistent with the wiring diagram.

Fuses

During the period 1934 to 1948 a 20-amp fuse was fitted in the battery earth lead, in a bakelite fuse holder (8032) screwed to the gear selector detent ball plug (7075) on the gearbox.

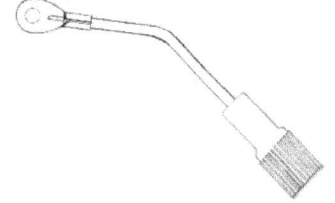

This is a reliable solution if regularly maintained. Note however that moisture entry can result in corrosion, leading to a poor earth connection.

Instrument lighting

Instrument lighting for machines numbered 1301 to 2550 consists of a single light bulb in an open socket between the ammeter and speedometer, illuminating both instruments.

Speedometer lighting for later machines is built into the instrument. Connections are shown on the applicable wiring diagrams.

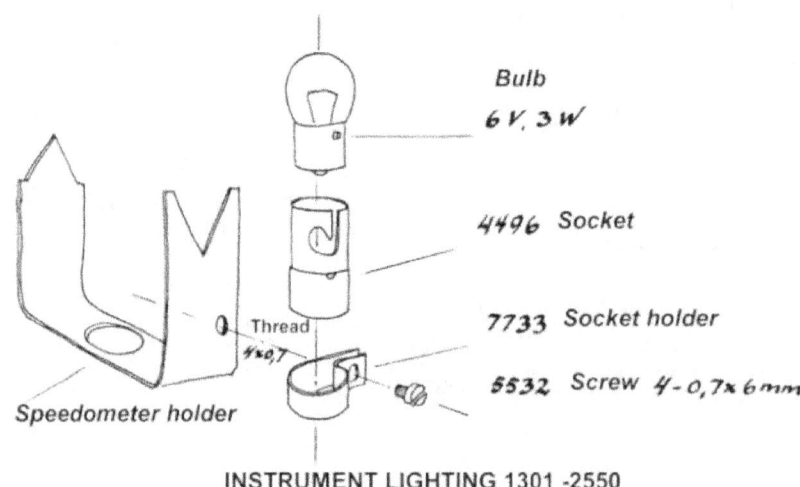

INSTRUMENT LIGHTING 1301 -2550

Gearbox

Maintenance

The gearbox of the Nimbus-C is very robust and does not need any regular maintenance, as lubrication is provided by the engine oil pump.

Repair

The most common reason for gearbox repair is when the machine develops a tendency to jump out of gear.

The early gearbox has three round dogs on each face of the layshaft second gear pinion (7704) which engage with holes in the face of either the layshaft 1st gear (7778) or 3rd gear (7777). Worn dogs or holes may result in unreliable gear engagement. Specialist welding to build up worn areas, followed by machining to the original profiles, can solve this problem.

If the hand-change or foot-change mechanism is in good order, without excessive slack, then the problem may lie with the gear selector shaft (7819 or 9210). The spring-loaded ball which acts on the detents of the selector shaft may, over time, wear the detents and the shaft such that the ball is no longer effective in locking the shaft in the exact position needed to correctly engage gears.

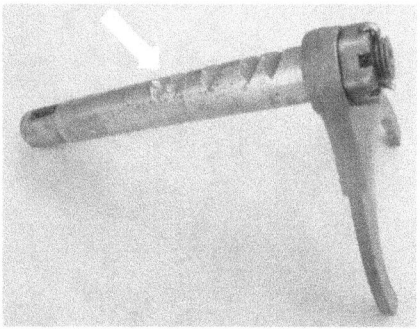

The shaft can be reconditioned. Worn areas can be built up by welding, and the profile of the detents and shaft restored by machining.

An alternative solution for foot-change machines is to exchange the early pattern selector shaft (7819), fork (7069) and linkage (9221) for a new later pattern combined selector shaft and fork (9210) and the corresponding linkage (9221).

In the later version of the gearbox the layshaft first gear bush (7472) may be damaged. This occurs only in the case of first gears with a hub length of 14mm fitted with a plain bush. Machining the hub length down to 11.5mm allows the improved shouldered bush (7472-2) to be fitted instead.

a symptom of wear, indicating that bearings require renewal.

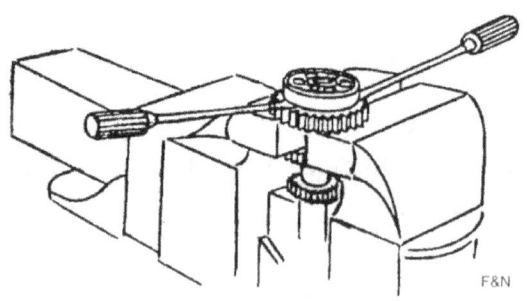

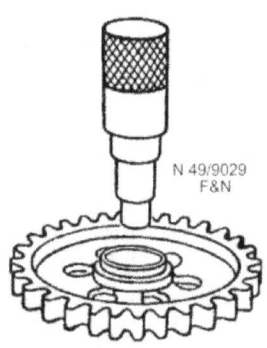

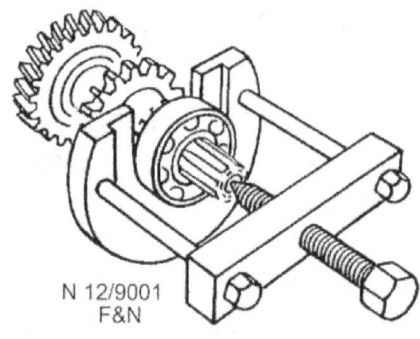

Noise from the gearbox can come from resonance (whining) or from the bearings (rumbling). Attempts to get rid of resonance have been made by modifying or buffing gear teeth, but this seldom results in a positive outcome. Noise coming from the bearings is of course

The gearbox and its end cover are made of aluminium and damage can be repaired by welding. Later gearbox castings feature a drilled lug which serves as the clutch cable abutment. This can fail. Repair consists of welding in place a suitable piece of aluminium, drilling a 10.5mm hole, and filing as required.

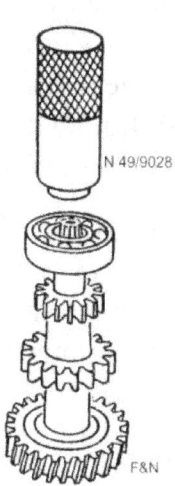

The drilling in the gearbox end cover for the clutch release rod (9215) may be worn into an oval shape, or the surrounding area of the cover may be damaged. This can be corrected by welding and machining.

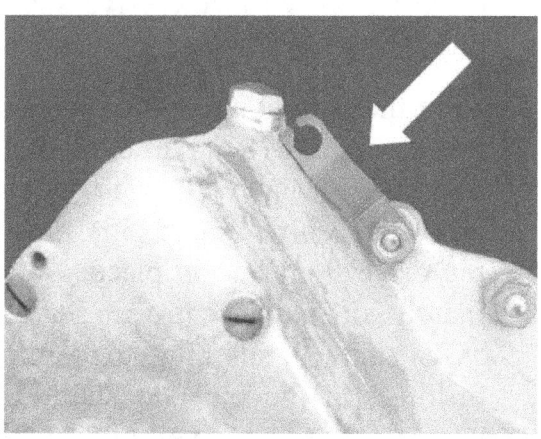

Repairs to the area of the oil feed hole in the gearbox housing can also be effected by aluminium welding and subsequent machining, but this work requires specialist skill.

Drive shaft

The drive shafts were not initially fitted with any shock absorber, but from frame number 2561, shock absorbing rubbers were incorporated. There are two versions, the first being used up to frame number 7091, and the second from from number 7092 on.

Dismantling

The original drive shaft, *without* shock absorber (7131), the so-called 'dog-bone', is fitted with a single compression spring (7828) bearing on the gearbox layshaft (7067). The spring can be renewed.

The two versions of the drive shaft *with* shock absorbers (8258 or

8258-2) are dismantled in the same way:

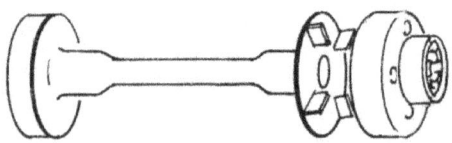

- Clamp the drive shaft in a vice and drive out the two 6 mm cross-pins (8257) using a drift and hammer. If a pin cannot be driven out in this way, it should be drilled through with a 5mm drill. The remainder of the pin can be heated with a small gas torch and allowed to cool, after which it should be readily driven out.

- Clamp the drive shaft tool N 27/9011 in a vice. Place the shaft in this tool, insert the puller and turn it clockwise so that it grips the splines of the shaft.

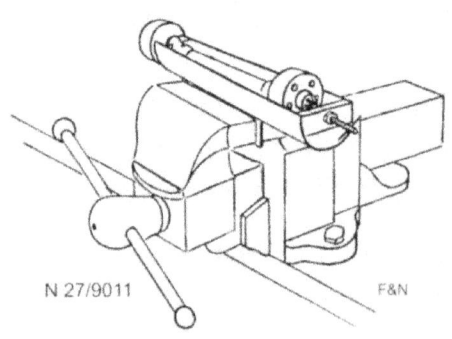

- Tighten the nut of the puller to remove the drive shaft hub (8270 or 8956) with the 9 balls (3846) and bolt (8255). It may be necessary to heat the intermediate shaft.

- This will release the 4 rubber blocks (8259) or 8 rubber blocks (8959), the shock absorber centre pin, and for the later version drive shaft, the containment ring (8958).

- Reverse the drive shaft in the tool and pull off the other hub.

Assembling

The drive shaft tool can be helpful in assembling the drive shaft by holding it in place, but the puller function will not be needed.

- Put 9 *new* ¼' greased steel balls (3846) in place in the drive shaft hub.

- Put the correct number of rubber blocks in place (8 of part 8259 or 4 of part 8959) and with a *new* shock absorber centre pin (8255) press the drive shaft hub in place. Note that the slot in the centre pin serves as a guide to aligning the 6mm drillings in the pin and shaft. If needed, use driveshaft tool

N12/9001. It is essential to apply some lubricant (for example rubber grease or soap) to the shock absorber rubbers, and to avoid using a sharp object when pressing them into place.

- When the 6mm drillings in the drive shaft and the centre pin are aligned, the cross-pin (8257) can be driven into place. Apply some anti-sieze grease to it before assembly - it may have to be removed at some future time.

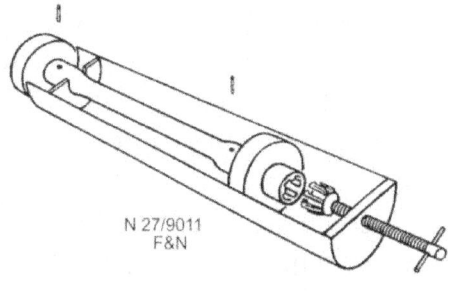

- Fit the other drive shaft hub in the same way.

Repairing the drive shaft

After taking the drive shaft apart, inspect the shock absorber centre pins (8255), the ¼' balls (3846), the drive shaft hubs (8270 or 8956) and all rubber blocks (8259 or 8959). Generally, the centre pin, balls and rubber blocks have to be renewed. Whether the drive shaft hubs have to be replaced, possibly by examples with less wear, depends upon the depth of the groove worn in the hub by the balls.

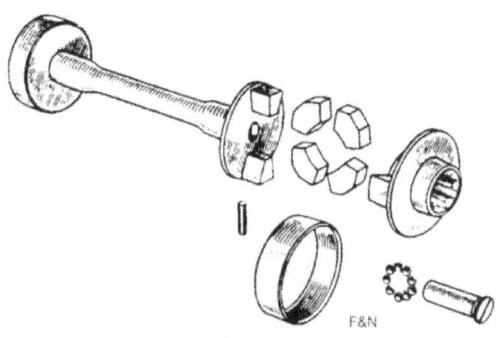

An alternative repair is to replace the shock absorber centre pin and balls arrangement by an after-market special one-piece pin ('kardanbolt').

Carburettor

Maintenance

The only required carburettor maintenance is cleaning. Contamination in the petrol and dust in the intake air can obstruct the jets and airways, leading to irregular running of the engine.

Dismantling and cleaning the carburettor in place

The carburettor can be dismantled for cleaning while still fitted to the engine. This will be the approach if there is a need for cleaning during a trip.

- Remove the float chamber base (7424). This will give access to the float (8553 or 8553-2) and for carburettors from 1938 to 1953, to the float needle (8554). Clean out the recesses in the float chamber base. Check that the jets are clear. Remove the accelerator pump base nut (8566) or bottom screw (8665). This will give access to the accelerator pump spring (8565) and piston (8564 or 9782). Clean the accelerator pump piston and bore. An interdental brush is useful for this. Remove the idle jet (7745) fitted to carburettors from 1934, and from 1938 to 1953. Use a small screwdriver to unscrew the jet. When the idle jet is free, it can be removed using the tip of a sharpened tooth pick. Always blow through the jets - never use a needle or sharp object. Needles are brittle and a broken piece will be very difficult to remove from a jet. If compressed air is not available, a single undamaged bristle from a steel brush may be useful. It is a suitable dimension, is flexible, and will not break as easily as a needle.

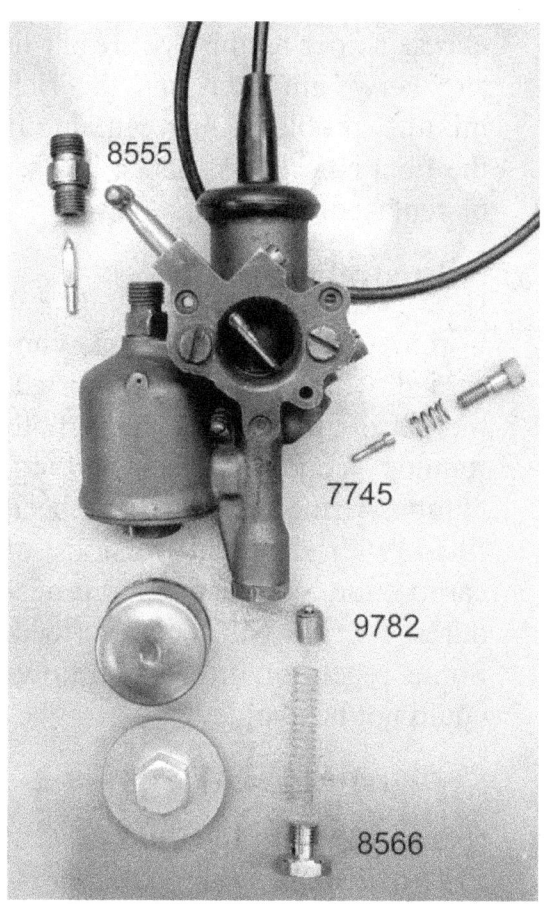

Wash the air filter (8584) in paraffin (kerosene). When dry, submerge it in thin oil then allow to drain.

After reassembly of the carburettor, it is absolutely essential to ensure that the float chamber base is securely tightened.

Repair

The needles (9884-2 and 10650) of the 1951-2 and 1953 carburettors are a press fit in the slide. Measure the distance between the tip of the needle and the bottom of the slide. This value must be between 28.5mm and 30mm. If the needle does not meet this requirement, it can be *carefully* tapped into place. *Make sure* the needle does not rotate.

Dismantling and cleaning the carburettor after removal

See Removing the carburettor.

When the carburettor has been removed, it will be easier to clear jets and internal passages of the carburettor housing.

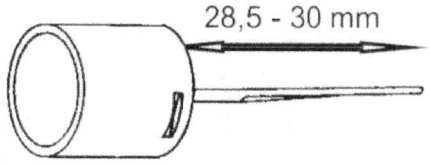

In the carburettor, as in other devices, it is between moving parts that wear occurs. As well as the slide and the mixing chamber bore, this includes the needle and the needle jet. Remove the carburettor needle and the jet, which may have worn to an oval shape. This can be verified by checking the fit of the needle in the needle jet. If wear is apparent, both parts should be renewed.

The needle jet of the 1950 and 1951 carburettors is pressed into the carburettor mixing chamber and normally cannot be replaced. It is however possible to convert these carburettors, so that a jet can be fitted as in the other models.

When the slide is no longer a close fit in the mixing chamber bore, excess air will be drawn past the slide resulting in a lean mixture and uneven running. The mixing chamber can be bored out and an oversize slide fitted.

Shake the float: if it has any leaks, it will contain petrol and have become effectively too heavy, leading to a rich mixture. Carefully melt the solder with a hot air gun or soldering iron. *Do not use an open flame! Take care, as excess internal pressure will damage the float..* Re-solder, and avoid adding excess solder as this will result in an overweight float and the rich mixture problem may remain. If the float has serious dents, it must be replaced.

Adjustment

Carburettor adjustment is concerned with ensuring the correct the petrol/air mixture, under both running conditions and idling. Nimbus-C motorcycles have been fitted with different versions of carburettor. Adjustment procedures for all will be described. Some repetition in what follows could not be avoided.

Carburettor 1934-1

(Frame numbers 1301 to 2487 and 2591 to 2516)

- The adjusting screw for the carburettor needle (7439) is fitted in a threaded bush in the slide (7676) and projects through a hole in the mixing chamber top (7413). Screw it all the way down and then back it off 2 to 3mm. The exact position has proven to be not very critical.

- The idle air screw (7746) is fitted at an angle behind the crankcase ventilation pipe (7684). Screw it all the way in. Make sure that the screw is really closed and that its travel is not limited by a fully compressed (overlength, non-original) spring. Contrary to the F&N 1934 technical leaflets, open the air screw 3 turns. The position of this screw has proven to be extremely critical as it heavily influences the mixture. When too open, the mixture will be too lean, resulting in overheated, white, sparking plugs. When too closed, the mixture will be too rich, resulting in soot formation in the combustion chamber and especially on the spark plugs, resulting in misfiring and eventually to total spark failure.

Remarks:
- These basic settings are for a 1934 to 1938 crankcase ventilation pipe with a 6mm hole to the outside air (1301 to 3863). This hole drains petrol in the case of carburettor flooding, but also functions as an extra air intake (this can be felt with a finger), resulting in a leaner mixture than the later version crankcase ventilation pipe which does not have the hole.

- Differences in sooting between the 4 spark plugs is not necessarily an indication of different performance of the respective cylinders.

- Further adjustment of both the needle adjustment screw (7439) and the air screw (7746) must be done by trial and error, observing the effect on the spark plugs.

- The throttle cable adjuster (7673) is fitted in front of the needle adjustment screw in the mixing chamber top. It is adjusted correctly when the slide just bottoms in the mixing chamber when the twist grip is fully closed.

- Start the engine: Turn on the fuel tap. Close the choke and operate the tickler until petrol flows from the float chamber vent. Turn on the ignition, open the throttle and operate the kickstart. Let the motor warm up until it has reached operating temperature and will run with the choke fully open.

- Close the throttle. If the motor stalls, raise the slide by means of the throttle cable adjuster in the mixing chamber top

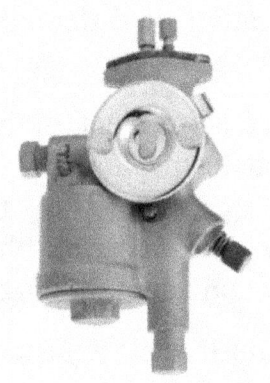

Carburettor 1934-2
(Frame numbers 2488 to 2500 and 2517 to 3863)

Adjustment steps 1 to 4 are as for carburettor 1934-1. Thereafter:

- Close the throttle. If the motor stalls, raise the slide by means of the throttle stop screw (5400) behind the air intake cover plate (7498).

Carburettor 1938
(Frame numbers 3864 to 8500)

- Fit the needle clip (8560) to the middle groove of the needle (8562) and secure it to the slide with the circlip (8561)

- The idle air screw (8569) is at an angle behind the crankcase breather (8593). Screw it home and then back it off 1½ turns.

- The throttle cable adjuster (8585) is in front of the needle adjustment screw in the mixing chamber top. Adjust it and the throttle stop screw (5400) behind the air filter until the slide just contacts when the twist grip is closed.

- Start the engine. Turn on the fuel tap. Close the choke (8597) by turning the lever forward. Turn on the ignition, open the throttle and operate the kickstart. Let the motor warm up until it has reached operating temperature and will run with the choke fully open (lever pointing left).

- Close the throttle. If the motor stalls, raise the slide by means of the throttle stop screw (5400) behind the air filter (8584).

- If the slide needs to be raised too far with the throttle stop screw, the engine needs a richer mixture. Lift the needle by moving the needle clip to the bottom needle groove.

- If the slide does not need to be raised at all, and the engine is running irregularly and with a sooty exhaust, the engine will need a leaner mixture. In that case, lower the needle by relocating the clip to the top groove.

Finally, adjust the idle air screw (8659) if necessary. Screwing it in richens the mixture.

Carburettor 1950
(Frame numbers 8501 to 9200)

This carburettor has no separate idle mixture adjustment. The carburettor needle (9778) differs from the needle fitted to carburettor type 1938 by having a single groove and being about 4 mm longer. It is fitted to the slide in such a way as to allow stepless adjustment by means of a threaded adjusting disc in the slide.

- Fit the carburettor needle clip (8560) to the single groove in the needle (9778) and fit both to the slide (9775) with the adjusting disc (9777).

- Screw the adjuster into the slide so that it is recessed by about ½ turn and is aligned with the stamped number 4 on the slide.

- Start the engine. Turn on the fuel tap. Close the choke (8597) by turning the lever forward. Turn on the ignition, open the throttle and operate the kickstart. Let the motor warm up until it has reached operating temperature and will run with the choke fully open (lever pointing left).

- Close the throttle. If the motor stalls, raise the slide by means of the throttle stop screw (5400) behind the air filter (8584).

- Next, adjust the carburettor needle position to get a correct mixture.

- If the mixture is too rich, the engine speed will increase when the crankcase breather (8593) is disconnected from the carburettor. Turn the adjusting disc to a lower figure - this lowers the needle.

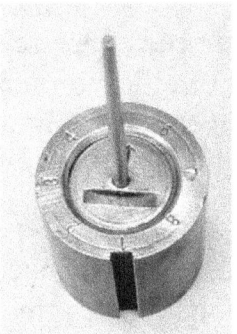

- If the mixture is correct, the engine stalls or will stop when the crankcase ventilation pipe is removed from the carburettor.

- After the mixture has been set as described, the idle speed can be adjusted with the throttle stop screw (5400) behind the air filter.

NB:
In practice (refer to MC-circular No. 108, 22/9/1950) satisfactory adjustment of this type of carburettor has been found difficult.

Carburettor 1951

(Frame numbers 9201 to 9630)
Adjustment steps 1 to 4 are as for carburettor 1950. Thereafter:

- To obtain the correct mixture, move the needle (round tapered needle 9884) by means of the adjusting screw (10102).

* If the mixture is too rich, the engine speed will increase when the crankcase breather (8593) is disconnected from the carburettor: Turn the adjuster anticlockwise to lower the needle.

* If the mixture is correct, the engine will stall or stop when the crankcase ventilation pipe is disconnected from the carburettor.

- Close the throttle. If the motor stalls, raise the slide by means of the throttle stop screw (5400) behind the air filter (8584).

- Final adjustment of the idle speed is done with the throttle stop screw.

NB:
In practice satisfactory adjustment of this type of carburettor has been found difficult.

Carburettor 1951-2

(Frame numbers 9631 to 11300)
Adjustment steps 1 to 4 are as for carburettor 1950. Thereafter:

- The face-ground needle (9884-2) is fitted into a slide with a vertical groove which engages with a guide screw in the carburettor body. Correct mixture setting is by means of the needle adjusting screw (10102).

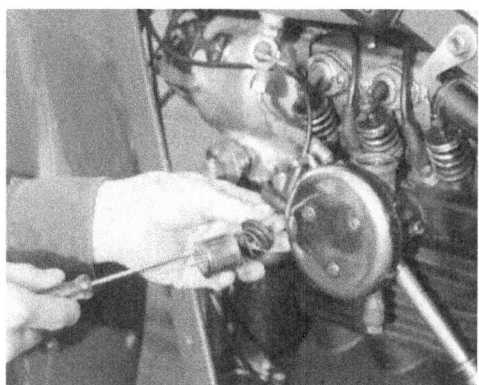

* If the mixture is too rich, the engine speed will increase when the crankcase breather (8593) is disconnected from the carburettor: Turn the adjuster anticlockwise to lower the needle.

* If the mixture is correct, the engine will stall or stop when the crankcase ventilation pipe is disconnected from the carburettor.

- Close the throttle. If the motor stalls, lift the slide by means of the throttle stop screw (5400) behind the air filter (8584).

- Final adjustment of the idle speed is done with the throttle stop screw.

NB:
In practice satisfactory adjustment of this type of carburettor has been found difficult.

Carburettor 1953
(Frame numbers 11301 to 14015)

- Fit the carburettor needle and holder (9884-2 or 10650) to the slide (9885), and using the needle adjusting screw (10102) position the holder so that it is recessed in the slide by about 1½ turns (or 1mm)

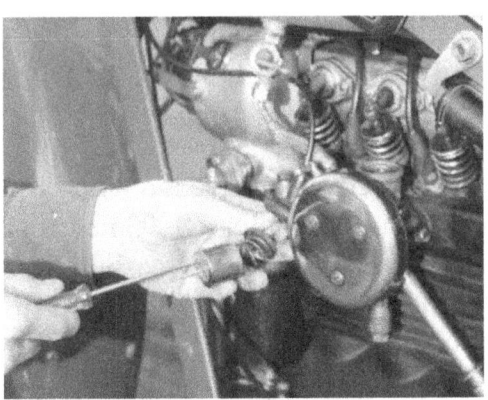

- The idle air screw (8569) is fitted at an angle behind the crankcase ventilation pipe (8592-2). Screw it home and then back it off 1½ turns.

- The throttle stop screw (8578) should be screwed out until the slide just bottoms in the mixing chamber. For the throttle stop to be effective, the throttlecable adjuster (8927) in the mixing chamber top (9886) must allow the slide full travel.

- Start the engine. Turn on the fuel tap. Close the choke (8597) by turning the lever forward. Turn on the ignition, open the throttle and operate the kickstart. Let the motor warm up until it has reached operating temperature and will run with the choke fully open (lever pointing left).

- * Adjust the throttle stop screw to give the lowest reliable idle speed.

- Test the initial setting of the idle air screw to see if any change results in smoother running.

- If idling improves as the air screw is temporarily screwed in, then the carburettor needle has to be lifted, to give a richer mixture.

- If idling improves as the air screw is temporarily screwed out, then the carburettor needle has to be lowered to give a leaner mixture.

- If any change to the idle air screw results in irregular running, then both the needle and idle air screw are correctly adjusted.

Note:
After any adjustment of the needle in the slide, the procedures from item, marked * have to be repeated.

Remarks on carburettor adjustment

The adjustment procedures described apply to carburettors and engines in good condition. *Carburettor adjustment alone cannot produce a satisfactorily running engine in the presence of worn carburettor jets, slide, or mixing chamber bore, incorrect ignition timing or valve clearances, or worn cylinders, rings, valves or valve guides.*

Main jet no.13 (8665) which is drilled to 1.3mm should not be used. Replace it by main jet no.15, which has a 1.5mm bore. A main jet bored out up to 1.8mm may be of benefit to an engine which has been rebored to 61.8mm and has a reconditioned (rebored) carburetter. Test all available jet sizes at operating temperature (marking the actual bore dimension on each), using full throttle in 2nd or 3rd gear.

Wheels

Nimbus-C wheels consist of a hub with an attached or an integral brake drum, 40 spokes with nipples, and a 19" x 3" rim (type WM3 x 19-40)

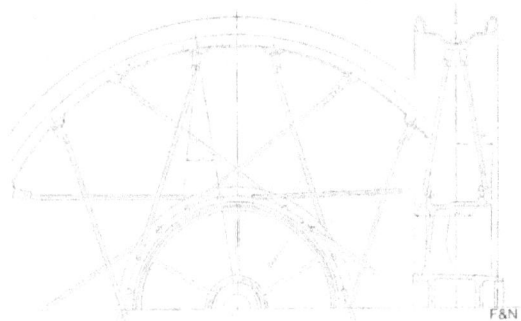

Hubs

The hubs incorporate machined seats into which the wheel bearings are a press fit. Over time a seat may become worn and the bearing may rotate within it, causing the hub to become unserviceable.

The left side bearing housing in the 180mm brake front wheel for example, often shows signs of such wear. It may be possible to eliminate the play by use of a bearing-grade anaerobic adhesive such as "Loctite". Alternatively, the hub can be repaired by sleeving the bearing seat.

The hubs have flanges with countersunk holes to accept the heads of the spokes. Spokes may break, or worse, may be pulled through the hole or damage the flange. The wheel hub is made of malleable cast iron, which is difficult to weld.

Spokes

Spokes for Nimbus-C motorcycles come in different lengths and different thread forms. Whether wheel-building or replacing individual spokes, take care to use only spokes of the correct specification. Note that from frame number 8501 the front wheel spokes (9578) have a slight bend at the threaded end.

Replacement spokes are now available in stainless steel.

Nipples

Like spokes, nipples come with different thread forms. Avoid any attempt to assemble non-matching spokes and nipples. Replacement nipples are available in various materials, including steel, nickel-plated brass, and stainless steel.

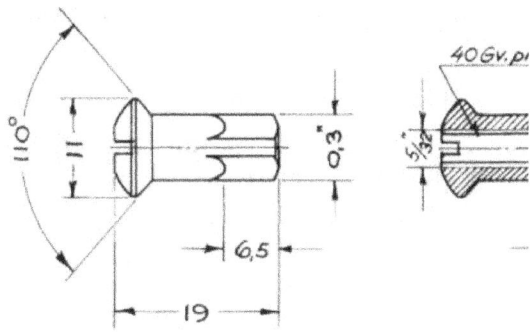

Rims

Over the years, it has been possible to buy wheel rims in different patterns and of different quality. In the period 1938 to 1945, "safety" rims were in use. These had a row of dimples in the bead seat to retain the tyre on the rim in the event of deflation. Chrome-plated wheel rims have also been supplied, in varying and sometimes poor quality. Replacement rims are also available in stainless steel.

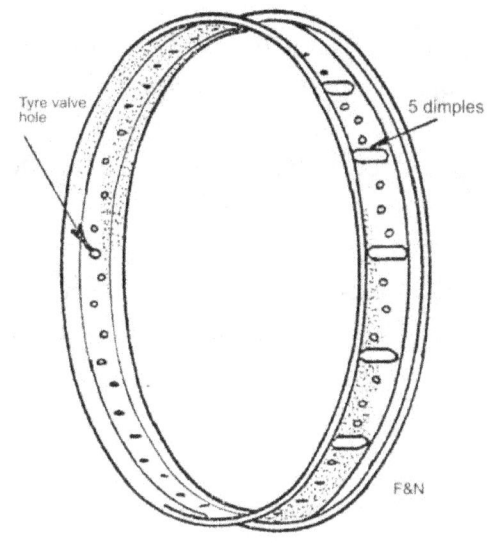

Checking wheel truth

A rotating wheel may sometimes appear to be 'out of round' because the flange of the rim may rise and fall as the wheel spins. As it is not the visible outer edge of the rim on which the tyre sits, but the bead seat of the rim (which cannot be seen when the tyre is in place), such a wheel may not be out of round at all. This can be checked, and if necessary corrected, with suitable wheel-building equipment.

Wheel-building and 'trueing'

Wheel-building - the art of lacing a spoked wheel is a specialist task. 'Inside' and 'outside' spokes of the correct gauge, length, and head bend are fitted to the wheel hub according to the lacing pattern for that particular rim. The lacing pattern determines how many spokes each individual spoke crosses, and into which hole in the rim it must be fitted.

Once assembled, the wheel must be 'trued'. A wheel is true when lateral (sideways) or radial (vertical) run-out is within the generally accepted tolerance range, that is, less than 1/16" or 2mm. Note that spokes are arranged at the rim in ten groups of four, two of which connect to each side of the hub. Run-out, lateral or vertical, is conveniently checked at ten points - the centres of each of these ten groups, and spoke tension is adjusted group by group.

'Trueing' involves first correcting lateral run-out (a sideways wobble as the wheel is spun) and thereafter correcting vertical run-out (the out-of-round condition, seen as a rise and fall of the rim). Wheels can be trued in place, but the work is easier if a wheel is removed and mounted in a wheel trueing jig, or a bench vice. Run-out can be measured with a dial gauge, or by some substitute arrangement.

To correct lateral run-out:
- Loosen those spokes in a run-out group which connect to the same side of the hub.
- Tighten those in the group which connect to the other side of the hub. This pulls the run-out area of the rim more to the centreline.
- Repeat the process until lateral run-out is reduced to a minimum.

To correct vertical run-out:
- Loosen all spokes in the group which is radially opposite a run-out group.
- Tighten all those in the run-out group. This pulls the run-out area of the rim closer to the hub.
- Repeat the process until vertical run-out is reduced to a minimum.

Finally, all spokes should be checked for tension by sharp taps with a metallic object. The 'ping' from each spoke should be similar.

Loose spokes are easily detected in this way and should be tightened.

Emergency repair

If one spoke has broken, there is a good chance that more will break soon. It is good practice, especially on a long trip, to replace any broken spoke as follows:

- Remove the wheel and take off the tire, tube and rim tape.

- Remove the broken spoke from the wheel rim and hub and unscrew the nipple.

- Take a suitable spoke from your spare parts kit (!) and fit it. The replacement spoke might have to be slightly bent during fitting, after which it should be straightened as well as can be done by hand. Once the nipple is fitted and tightened, the straightening process should be complete.

If the threaded end of the spoke protudes through the nipple the excess length must be cut or filed away to avoid tube damage.

- Fit the rim tape, tube and tire.

- Fit the wheel.

- Finally, be sure to check the tension of all spokes, as one broken spoke may be the result of other spokes being loose.

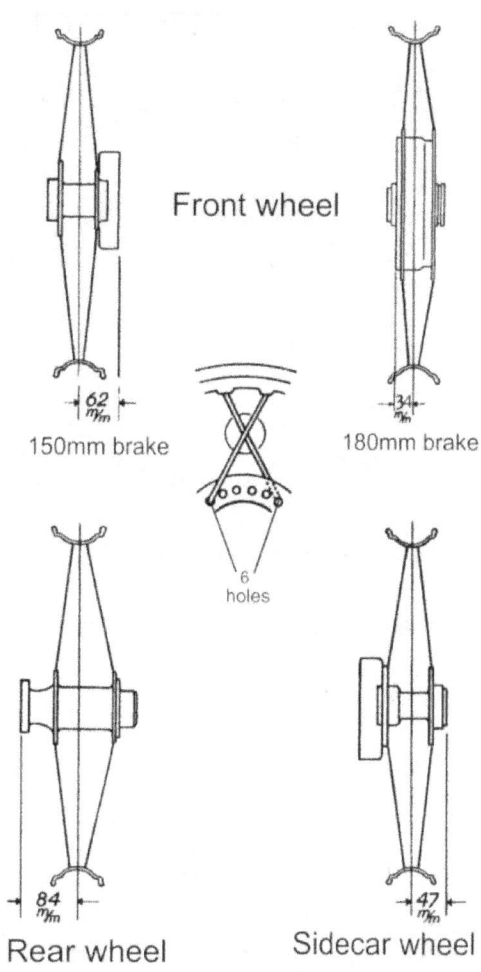

Front wheel

Removing the front wheel

Put the motorcycle on the centre stand with a 20mm board under the stand.

Front wheel with 150mm brake:

- Disconnect the front brake cable (7671 or 8248) from the brake lever on the handlebar.

- Remove the speedometer cable (7753) from the speedometer drive (7153 or 8359).

- Loosen the 2 wheel spindle nuts (7240).

- Remove the front wheel by pulling it down out of the forks. If necessary, use a soft faced hammer to knock the spindle free.

Front wheel with 180mm brake:

- Release the front brake cable (8707 or 8945) from the brake cam arm on the brake plate.

- Disconnect the speedometer cable (9903) from the brake plate.

- Remove the right-side spindle nut (7240 or 9480).

- Slacken the left side spindle clamping nut (7259).

- Pull out the wheel spindle (8677 or 9279) using a tommy bar (a bolt or screwdriver will do) through the 12mm hole in the spindle head. If the spindle is difficult to remove, drive it out from the right, using a drift which will not damage the end of the spindle.

- Remove the front wheel from the forks.

Fitting the front wheel

Put the motorcycle on the centre stand with a 20mm board under the stand.

Front wheel with 150mm brake

The wheel must be fully assembled before fitting: spindle, bearings, brake plate and shoes, speedometer gearbox complete with worm gear (7372 or 8361), and brake cable must all be in place. (The speedometer cable is attached later).

- Place the wheel between the forks. Lift the front wheel spindle into place. If necessary, use a soft faced hammer to knock the spindle home.

- Tighten the two spindle nuts (7240)

- Connect the brake cable (7671 or 8248) to the handlebar brake lever and adjust it as needed.

- Attach the speedometer cable (7753) to the speedometer gearbox (7153 or 8359) and spin the wheel to check its functioning.

Front wheel with 180mm brake:

The wheel must be fully assembled before fitting: bearings, spacer and worm gear (7372 or 8361), and brake plate and shoes must all be in place.

- Place the wheel between the forks. Fit the dust cover (8864 or 8864-2) on the left side of the brake drum. Note that the front wheel for the '39 fork has a hub cover (8678) instead.

- Slide the wheel spindle (8677 or 9279) through the left fork leg, hub, brake plate and right fork leg.

- Fit the spindle nut (7240 or 9480) and tighten it, using a tommy bar (a bolt or screwdriver will do) through the 12mm hole in the spindle head to prevent it turning. *Take care*: Do not yet tighten the clamping nut on the left, as the forks may not be aligned at this stage.

- Test that the forks move freely throughout their travel.

- Now tighten the spindle clamping nut (7259) and check again that the forks are not binding and move freely.

- Fit the brake cable (8707 or 8945) to the brake cam arm and adjust the cable as needed.

- Fit the speedometer cable (9329 or 9903) and spin the wheel to check its functioning.

Dismantling the front wheel

- Remove the wheel and set aside the brake plate and the separate speedometer gearbox fitted to wheels with the 150mm brake.

- The worm gear (7372 or 8361) has a left-hand thread - it must be removed by turning it clockwise. Use C-spanner N 52/9031 or apply a punch to the worm gear holes.

- Remove the threaded hub seal (8678) if fitted. Use C-spanner N 52/9031 or a drift.

- Remove the wheel bearings (7089 or 9477-2)
On machines with the early (internal slider) forks and wheel bearings 7089/SKF 6204, use tool N 29/9033.

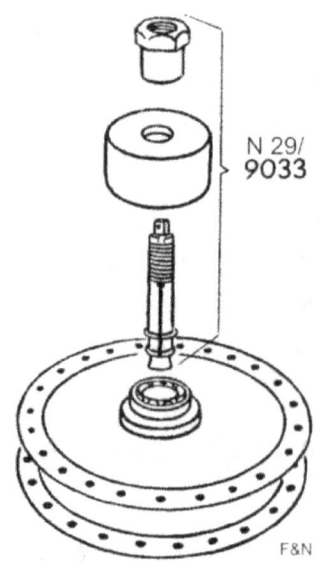

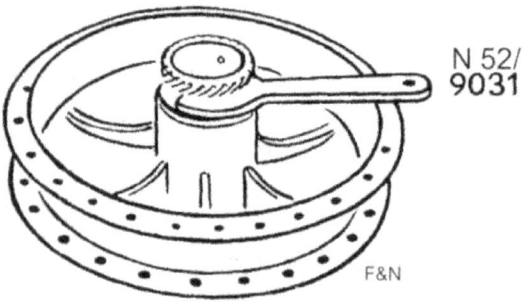

For all machines with the later (external slider) forks, the front wheel bearings (9477-2/SKF 6303) must be driven out using tool N 55.

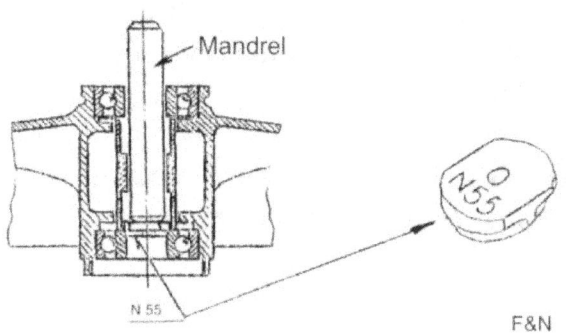

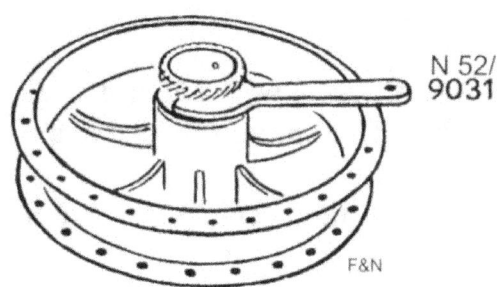

52/9031 or apply a punch to the worm gear holes.

If these special tools are not available, use a long flat-faced punch to gradually drive out the bearing on the opposite side. Work evenly around the edge of the inner bearing race so as to avoid tilting (and locking) the bearing in the hub.

Assembling the front wheel

Drive the brake-side wheel bearing (7089/SKF 6204 or 9477-2/SKF 6303) into place using tool N 53/9032, or using a punch on the outer race only, working evenly around the circumference to avoid tilting the bearing in the hub

- Fit the worm gear (7372 or 8361) which has a *left-hand* thread and has therefore to be screwed in by turning it anti-clockwise. Use C-spanner N

For a front wheel with 180mm brake, put the distance piece (8679 or 9479) in place.

- Drive the other bearing into the hub. On the early (internal slider) fork machines, use bearing SKF 6204-2RS (formerly named 6204 Z) and for all later (external slider) fork machines use SKF 6303-2RS (formerly named SKF 6303 Z), which is a sealed bearing.

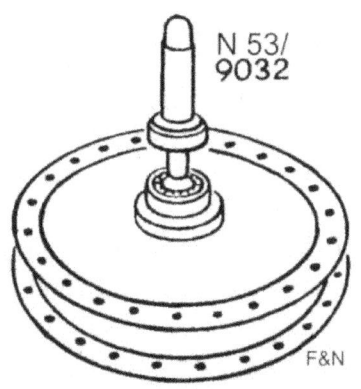

Speedometer gearbox:
removal, dismantling and reassembly

Front wheel with 150mm brake and speedometer gearbox ratio 1:3
(Frame numbers 1301 – 2900)

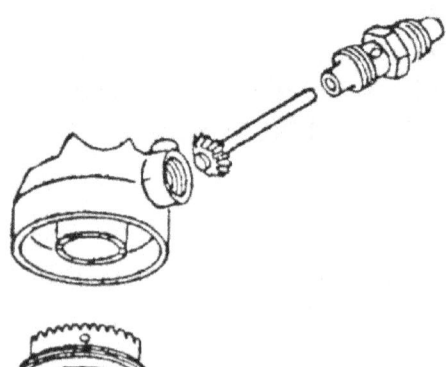

Remove the front wheel and detach the speedometer gearbox housing (7153) from the front wheel spindle.

- Remove the speedometer cable adapter (7374) and speedometer pinion (7373).

- The worm gear (7372) has a *left-hand* thread - it must be removed by turning it clockwise. Use C-spanner N 52/9031 or apply a punch to the worm gear holes. Reassembly is done in the reverse order.

Front wheel with 150mm brake and speedometer gear ratio of 1:2
(Frame numbers 2901 – 4601)

Remove the front wheel and detach the speedometer gearbox housing (8359) from the front wheel spindle.

- Remove the speedometer cable adapter (8362) and speedometer pinion (8360).

- The worm gear (8361) has a *left-hand* thread - it must be removed by turning it clockwise. Use C-spanner N 52/9031 or apply a punch to the worm gear holes. Reassembly is done in the reverse order.

Front wheel with 180mm brake

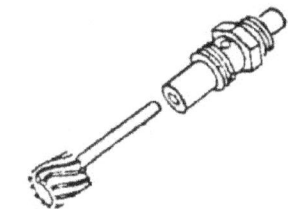

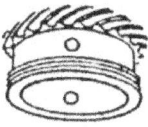

Remove the front wheel and take the brake plate (8676 or 9267) from the front wheel spindle.

- Remove the speedometer cable adapter (8362) and speedometer pinion (8360).

- The worm gear (8361) has a *left-hand* thread - it must be removed by turning it clockwise. Use C-spanner N 52/9031 or apply a punch to the worm gear holes. Reassembly is done in the reverse order.

Rear wheel

Removing the rear wheel

Place the motorcycle on the centre stand and remove the split pin (3865) and washer from the brake pull rod (7674 or 8498). Remove the brake pull rod from the brake cam arm (7124 or 8380).

- Remove at the right side, the single fixing bolt (7189) securing the brake plate to the frame, and at the left side the two fixing bolts (for frames prior to No 2400: one bolt) of the final drive housing.

- Loosen the 2 wheel spindle nuts (7240) leaving them in place on the spindle.

- Loosen the nut (7260) of the pillion seat and rear mudguard pivot bolt.

- Swing the mudguard and pillion seat clear of the rear wheel and secure them in this position using a strap from the passenger grab handle to the steering damper.

- Place supporting blocks under the frame or rear foot rests.

- Withdraw the rear wheel from the frame. *Take care:* if the drive shaft (7131 or 8258-2) is also withdrawn with the wheel, the small compression spring housed in the front transmission shock absorber will drop out and must be recovered.

Dismantling the rear wheel

Remove the wheel, as detailed above. To facilitate working on it, the wheel can be held horizontally in a bench vice by one of the spindle nuts (7240).

- Clamp the rear wheel in a vice with the final drive housing (7121) down and remove the brake side spindle nut.

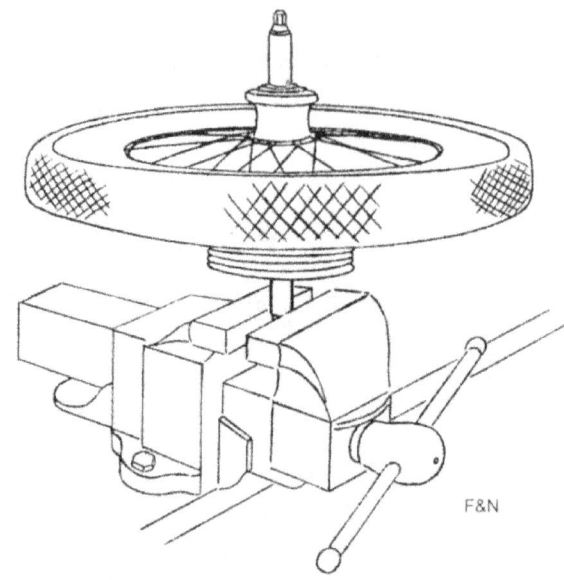

- Lift the brake plate (7123, 8375 or 9546) with brake shoes, cam, and brake cam arm away from the spindle.

- Flatten the locking tabs (8385) to allow removal of the 8 bolts (8384) securing the brake drum. Detach the brake drum. If necessary, tap sharply around the rim of the brake drum using a soft faced hammer. *Note:* The brake drum of the rear wheel for frame numbers 1301 – 3000 is riveted onto the rear hub and cannot be removed.

- Replace the brake side spindle nut and turn the wheel over in the vice so that the final drive housing is now uppermost. Remove the final drive side spindle nut (7240).

- Remove the 9 bolts (7191) which secure the final drive cover (7122, 8223 or 8387) and remove it. If necessary, tap sharply around the edge of the cover using a soft faced hammer.

- Flatten the locking tabs (8108) of the 8 bolts (7190) securing the crown wheel (7845, 8058, 8390 or 8391) and remove the bolts. The crown wheel, final drive housing and pinion can now be removed together.

- Drive out the final drive side spindle cross-pin (7193) - near the wheel bearing - using a thin punch and hammer. *Take care* not to damage the shims.

- Lift the rear wheel free of the spindle. The inner race of the thrust ball bearing (7134 / SKF 51205, frame numbers 1301 to 3000) or the inner race and roller assembly of tapered roller bearing (8381/ SKF 303040, frame numbers 3001 on) will come off with the wheel. These are loose and must be set aside for storage before the wheel is stood upright.

- Release the vice and take out the spindle with nut, brake side cross-pin, dust cover (8378), and brake side inner race and roller assembly (or ball bearing). Remove the remaining fitted cross-pin and other parts from the spindle.

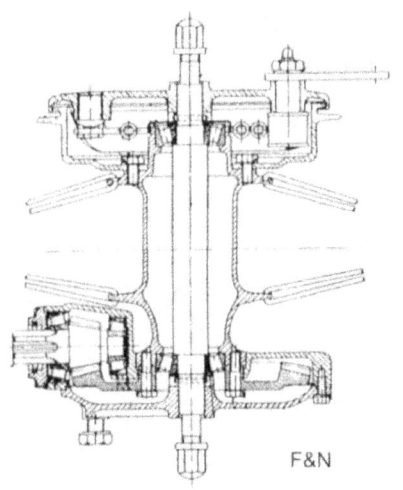
F&N

- Remove the tapered roller outer races from the hub. Use Nimbus special tool N 29/9013 with 52 mm pull ring, or a similar bearing puller. If neither is available, use a long flat-faced punch.

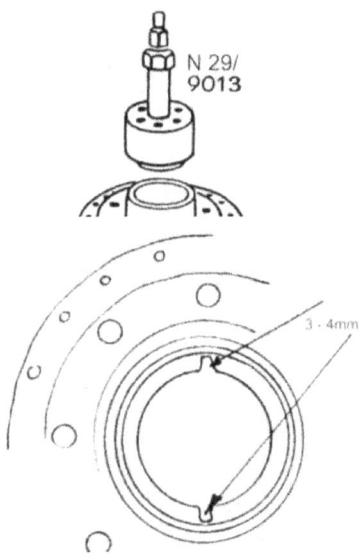

It may be helpful for any later bearing removal, to file a groove at both sides of the hub, to provide better engagement of a punch.

Initial assembly of the rear wheel and adjustment of bearings

Rear wheel with 150 mm brake

(Frame numbers 1301 to 3000):

These wheels are fitted with deep groove ball bearings (7088/ SKF 6304) at both ends of the hub and at the final drive side with the thrust ball bearing (7134/SKF 905). This is no longer available: bearing SKF 51205 can be used instead. For adjustment purposes, bearings should be dry (not greased). They should be grease-packed only when adjustment is complete.

- Fit the right-side ball bearing, cross-pin (7193), special distance piece (tool N 32L/9014-2) and the spindle nut (7240) to the wheel spindle. Clamp the spindle vertically in a vice, by the spindle nut.

- Lower the rear hub (or the wheel if assembled, and tyre if fitted) over the wheel spindle with the brake drum down.

- Fit the thrust ball bearing into the hub. *Take care:* The inner diameters of the two bearing races are different. The bearing race with the larger inner diameter, 25.2 (the original bearing SKF 905 has a diameter of 27 mm) has to face the brake side.

- Place the inner race of an old ball bearing (SKF 6304), then special distance piece (tool N 32L/9014-2) and the spindle nut (7240) to the wheel spindle. Tighten the whole assembly.

- Check whether there is clearance between the thrust bearing and the inner race. If this is the case, the clearance has to be eliminated with shims (7577 or 8274). These are the same type as used for the pinion wheel adjustment. Place the shims as required between the *right-hand* bearing (at the brake side) and the wheel hub until all play has been eliminated and the assembly is under just slight stress. The bearings, when dry, must run smoothly when the wheel is spun.

- Bearings should be well-greased and therefore ready for service once adjustment is complete.

Source: *Andersen, J. B. (1996) Nimbus- model C 1934*

Rear wheel with 180 mm brake (Frame numbers 3001 to 14015).

The wheel should be fitted with either new tapered roller bearings (SKF 30304) or with serviceable used bearings which have been thoroughly cleaned. For adjustment purposes, bearings should be dry (not greased). They should be grease-packed only when adjustment is complete.

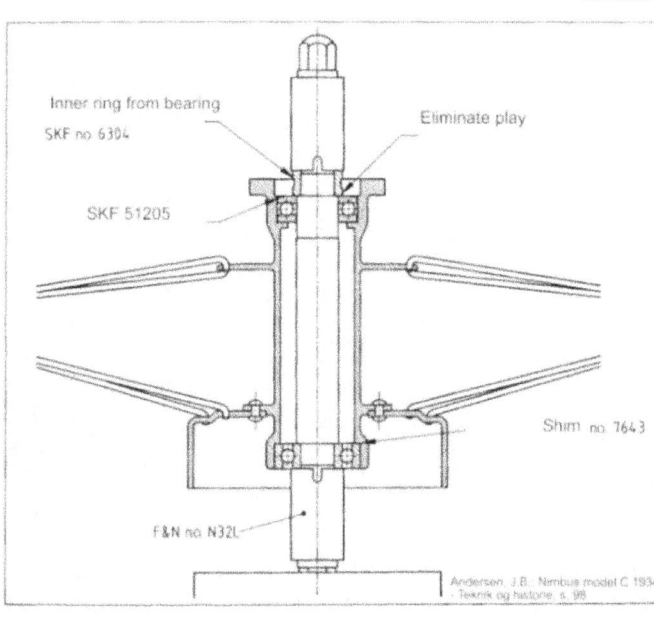

- Fit an outer bearing race in the bearing seat at each end of the hub, using special tool N53/9032. If a race is not a press fit in the seat (that is, if it does not need to be driven into place) then it ought be secured with bearing-grade "Loctite". If this is not feasible, the rear wheel hub will need repair.

- Fit a dry inner bearing race, dust cover (8378) and cross-pin (7193) to the spindle.

- Fit special tool N32L/9014-2 and a spindle nut (7240) to the same end of the rear wheel spindle and clamp the spindle vertically in a vice, by the spindle nut.

- Lower the rear hub (or the wheel if assembled, and tyre if fitted) over the wheel spindle with the brake drum down.

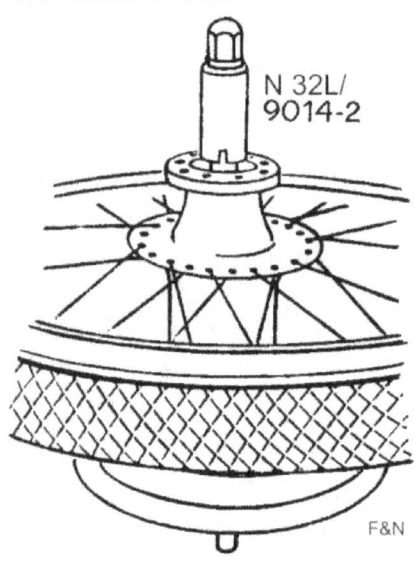

- Slide the second inner race into place, then fit the special distance piece (tool N 32L/9014-2) and the spindle nut (7240) to the wheel spindle. Tighten the whole assembly.

- Put shims (8113 and/or 8277) as required under the inner bearing race until all clearance has been eliminated and a *little* tension is reached. The tapered rollers bearing should run freely and evenly when the wheel is spun.
- Bearings should be well-greased and therefore ready for service once adjustment is complete.

Final drive pinion

Removing the pinion bearings

Remove the pinion cover (8333) and pull the pinion out of the final drive housing. If the outer race of the large pinion bearing is tight, clamp the pinion in a vice by its splines and tap sharply on its housing with a soft faced hammer.

- Remove the small ball bearing (7133/SKF 6302) or the inner race of the tapered roller bearing (8383/SKF 30302). From the nose of the pinion. Mount the pinion in the special drive shaft tool N27/9011 and release the bearing by using a punch and hammer. If this special tool is not available, use a small bearing puller.

- Remove the inner race of the large pinion bearing with a punch and a hammer. It will be difficult to engage a punch on the bearing race of the 14-tooth solo-gear pinion. A sharp chisel can be used instead (e.g. a woodworking chisel, which will be badly damaged) or the bearing can be cut away using a cold chisel.

Remove the inner race of the small tapered roller bearing using special puller N29/9013 or by heating the final drive housing then striking the housing down onto a wooden block.

Fitting and adjusting the pinion bearings

Fit the bearings to the pinion: For pinions 1301 to 3000 a tapered roller bearing (7132/SKF 30205) and a ball bearing (7133/SKF 6302) are used. For pinions 3001 to 14015 two tapered roller bearings are used. Bearings should be assembled dry (without grease).

First fit the inner race of the large bearing (7132/SKF 30205) on the pinion. Use hollow drift N49/9028.

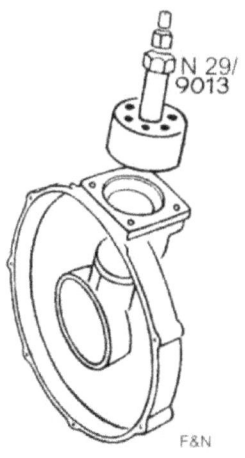

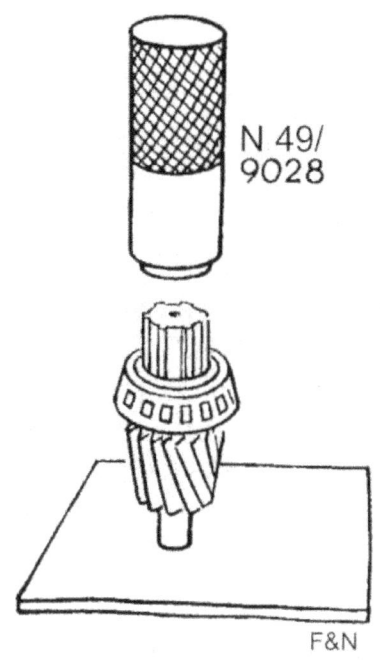

Then, fit the ball bearing (7133/SKF 6302) or the inner race of the small tapered roller bearing (8383/SKF 30302) on the pinion.

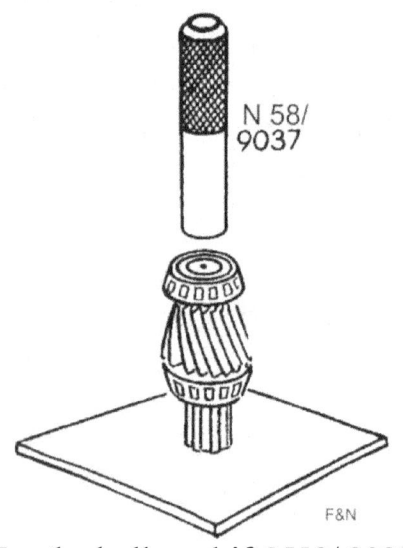

Use the hollow drift N58/ 9037).

For motorcycles with frame numbers 3001 to 14015 fit the outer race of the small tapered roller bearing (8383/SKF 30302). Start by putting 3 shims (8021) under the race. Drive the race home with hollow drift N58/9037. If the bearing race does not need to be driven into place and can be removed without a puller, it should be secured *after* the final adjustment with bearing-grade Loctite.

Adjust the pinion to wheel spindle distance:
- Fit the pinion into the final drive housing.

- Fit the outer race of the large pinion bearing (7132/SKF 30205).

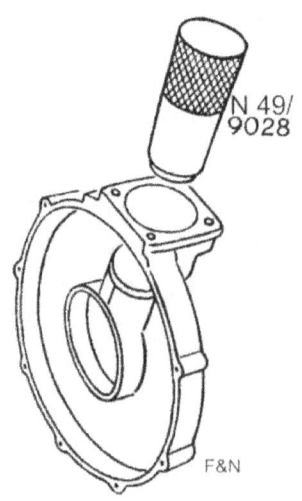

- Fit a new cork seal (8392) into the pinion cover (8333).

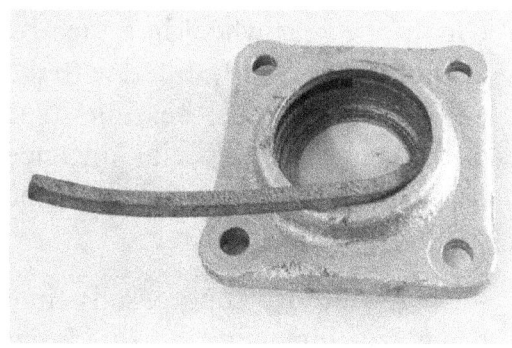

- Fit the cover (7128 or 8333) with the shims (7577 and/or 8274) and its 4 securing bolts (7191). The shim package shall be such that after the fitted pinion has been driven home by a *sharp tap* with a hammer, it can be rotated by hand, without detectable play.

- Fit the fixed part of the pinion adjustment gauge (tool N 28/9012) onto the final drive housing. Hold the housing in a vice.

- Check with the special double-ended 'Go/No go' gauge (part of tool N28/9012) that the rear face of the pinion clears the gauge tip marked 'GAA' (Go) when applying a light pressure and touches the gauge tip at the side 'IKKE GAA' (No go). If the pinion turns freely under both tests, a 0.1mm shim has to be removed from under the outer race of the small pinion bearing. If the pinion touches the gauge under both tests a 0.1mm shim must be added.

If this special tool is not available, measure the distance from the smaller face of the pinion to the opposite inside edge of the final drive housing. This should be 106.0 mm for an original pinion and 106.7 mm for a pinion with helical teeth.

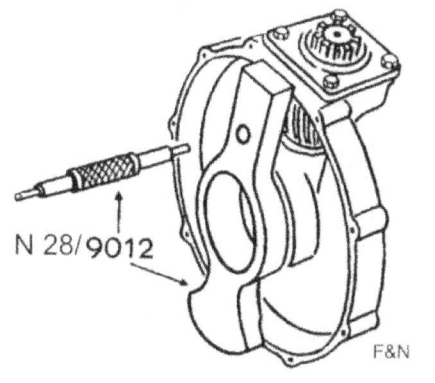

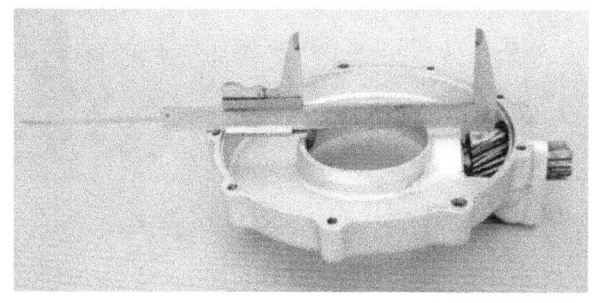

- Extract the inner race of the small pinion bearing with a special puller N29/9013) or knock it out after heating the final drive housing.

- If the inner race of the small pinion bearing does not need to be driven home or does not need to be extracted with a puller but can be removed by hand, then the bearing must be secured with bearing-grade Loctite *after* the final adjustment.

- Finally adjust the number of shims under the pinion cover. Grease both bearings and fit the cover with 4 bolts (7191) and spring washers. If if sidecar-gearing is fitted attach the indicating tab stamped "2:59"

Crown wheel and pinion

Adjusting crown wheel and pinion

Clamp the rear wheel in a vice by a spindle nut (7240), fitted with the adjusted rear wheel bearings (see above) and with special distance piece, tool N32L/9014-2, with the brake side brake down.

- Lower the dry (ungreased) final drive housing, with the fitted and adjusted pinion (see above) over the spindle.

- Fix the crown wheel to the hub using 2 of the 8 bolts (7190) without locking tabs. *Make sure* that the crown wheel and pinion teeth are engaged before the bolts are tightened.

- Fix the crown wheel cover (7122 etc.) using 3 of the 9 bolts (7191).

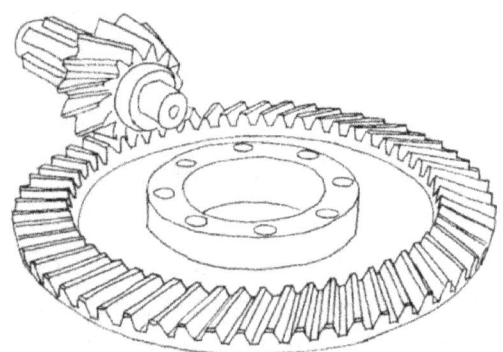
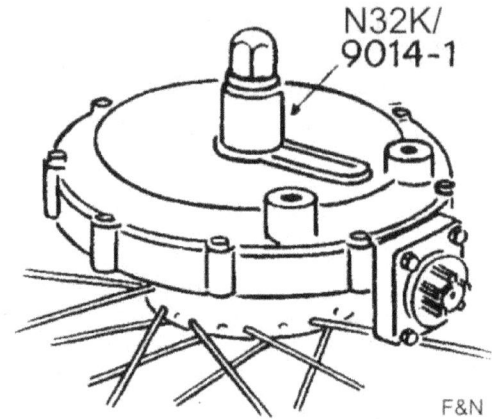

- Fit the *short* distance piece, special tool N32K/9014-1, and the second spindle nut (7240) and tighten the hub-final drive assembly.

- There will be clearance between the teeth of the crown wheel and the pinion.

With a *new crown wheel and pinion* this clearance will be excessive and it must be reduced to 0.2mm. Adjustment of the clearance is by means of shims fitted between the pinion outer bearing and the pinion cover, first determining the shim set needed to *just* eliminate play. This arrangement of shims (8113, 0.15mm) and 8277, 0.05mm) must found by trial, each time tightening the pinion cover and spinning the wheel *clockwise*. (This is opposite to the direction of rotation when riding but simulates what happens when the engine drives the rear wheel). When zero play condition is reached, *removal* of shims totalling 0.25mm will produce correct adjustment.

For a *used crown wheel and pinion* first of all make sure that the pinion teeth do not 'bottom' in the crown wheel. Should this be the case it will not be possible to obtain the correct clearance.

- Take the whole assembly apart, including the rear wheel bearings, and grease the bearings. Pack 250 grams of grease in the final drive housing and grease the pinion bearings.

- Fit the cross-pin, dust cover, lubricated bearing and a spindle nut (7240) to the brake side of the wheel spindle. Clamp the assembly by the spindle nut in a vice. Lower the rear wheel over the spindle, with the brake side down. Put the shim set for the rear wheel bearing *under* the bearing on crown wheel side, put the shim set for the crown wheel *above* the bearing and fit the cross-pin.

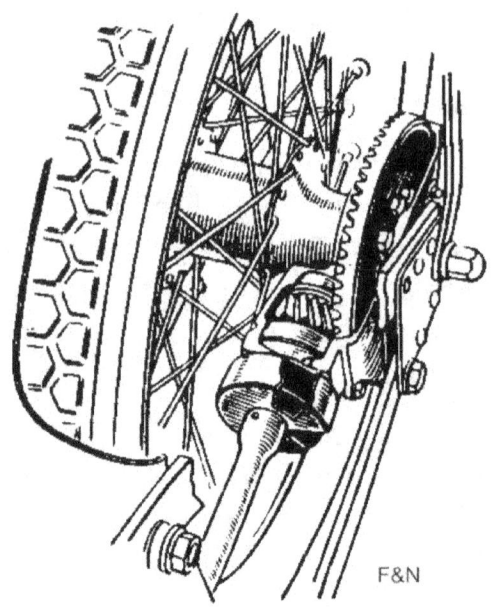

- Put the final drive housing and the crown wheel in place. Make sure that the teeth of the crown wheel and pinion are engaged, then fit the crown wheel locking tabs and 8 bolts (7190). Tighten the bolts and bend up the locking tabs *Take care:* Do *not* use a grease nipple for either greasing the drive shaft housing, or for the cover. In the worst case, the grease nipple ball can be dislodged causing damage to the final drive gears. Instead, blank off the grease nipple hole with a bolt (7749) and lubricate the final drive by removing a frame bolt (7189) and applying grease by grease-gun through the hole in the frame and crown wheel cover.

- Fit the crown wheel cover (7122) and its 9 bolts (7191) and spring washers.

Final assembly of the rear wheel (Frame numbers 3001 to 14015)

- Fit the brake drum (8370-2) with its locking tabs and 8 bolts (8384) to the rear hub. Bend up the locking tabs after tightening the bolts.

- Fit the brake plate (with brake shoes, springs, cam and cam arm)

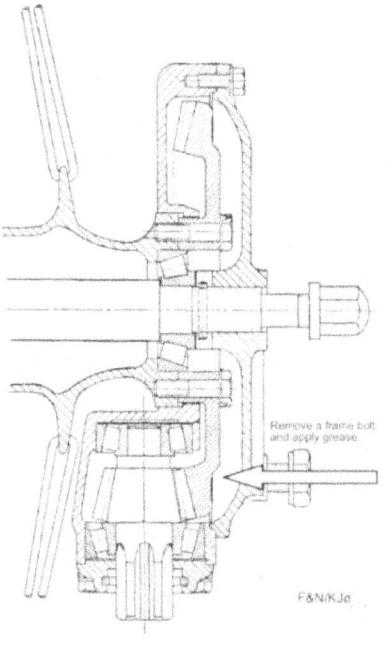

to the rear wheel spindle. (See brakes).

Fitting the rear wheel

- Check that the small compression spring (8388) is correctly positioned in the rear of the lay shaft and fit the drive shaft onto the lay shaft splines.

- Place the rear wheel into the frame. *Make sure* that the brake plate is correctly fitted over the cross-pin in the wheel spindle. The crown wheel cover and the brake plate castings incorporate ridges which engage with the frame ensuring correct hub orientation. The drive shaft and final drive pinion splines must be aligned as the wheel is moved into place.

- Fit the 2 or 3 rear wheel fixing bolts (7189) and spring washers, and hand tighten them. *Be careful* not to confuse the engine mounting bolts (7269, length 25mm) and the rear wheel fixing bolts (7189, length 32mm).

- Lower the rear mudguard and pillion seat and guide the two slotted mudguard stays over the wheel spindle.

- Tighten all rear wheel fixing bolts (7189) and both spindle nuts (7240). Tighten the nut (7260) of the mudguard pivot bolt (7816 or 9646)

- Fit the brake pull rod (7684 or 8498) with washer and split pin (3865).

Brakes
The 150 mm brakes

Dismantling the front brake

Remove the front wheel (see the applicable section).

- Remove the spindle nut (7240) and take the brake plate from the wheel spindle.

- Remove the two brake shoe return springs (7559), fitted between the brake shoes (7154) and the brake shoe pivots (7137).

- Pull the securing pins from the brake shoe pivots on the brake plate.

- Remove the brake shoes.

- Remove the split pin (3866) from the brake cable pin (7479) and remove the cable (7671 or 8248).

- Take the brake cam arm (7152-2 or 7152) and return spring (7560 or 8247) from the brake plate.

- Remove the split pin (7138) roller pivot (7163) and roller (7130) from each brake shoe.

Assembling is done in the reverse order.

Note:
It is important to install the brake shoes so that the linings are "leading" and the pivot ends of the shoes "trailing" when the wheel is rotating forward: there will then be an appreciable 'servo-effect'.

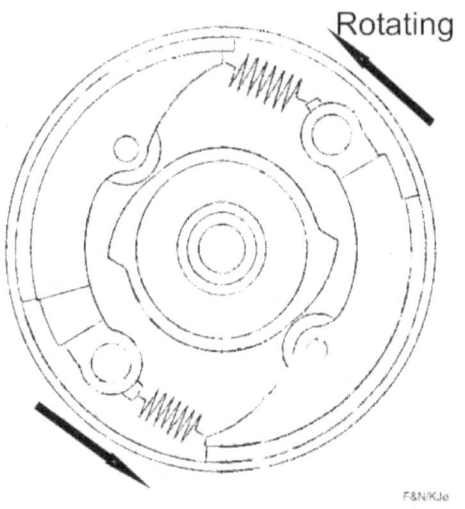

Dismantling the rear brake

- Remove the rear wheel (see the applicable section).

- Remove the spindle nut (7240) and take the brake plate (7123) from the wheel spindle.

- Remove the two brake shoe return springs (7559) fitted between the brake shoes (7154) and the brake shoe pivots (7137).

- Pull the securing pins from the brake shoe pivots on the brake plate.

- Remove the brake shoes
.
- Remove the split pin (3866) from the brake cable pin (7479) and remove the cable (7671 or 8248).

- Take the brake cam arm (7152-2 or 7152) and return spring (7560 or 8247) from the brake plate

- Remove the split pin (7138) roller pivot (7163) and roller (7130) from each brake shoe.

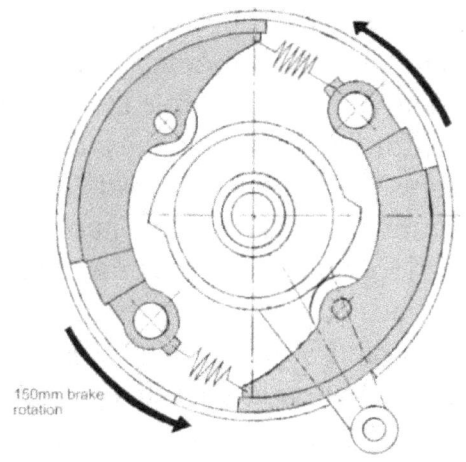

Assembling is done in the reverse order.

It is important to install the brake shoes so that the linings are "leading" and the pivot ends of the shoes "trailing" when the wheel is rotating forward: there will then be a useful servo-effect.

The 180 mm brakes

Dismantling the brake

- Remove the front or rear wheel (see the applicable section).

- Remove the spindle nut (7240) and take the brake plate (8676 or 9267 for front wheel and 8375 or 9546 for rear wheel) from the wheel spindle.

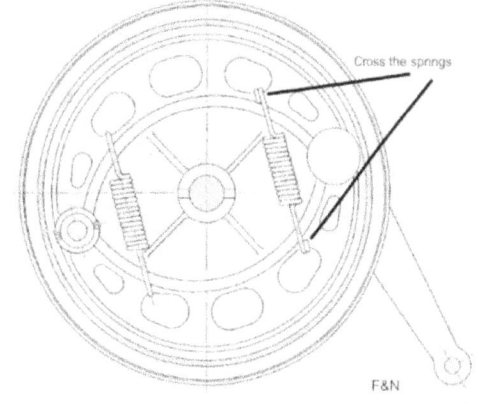

- Remove the 3 brake return springs (8379) from the brake shoes (8372 or 9275).

- Remove the brake shoes. Brake shoes (8372) with a pivot eye are secured with a split pin (4518), which has to be removed first.

- Remove the nut (8683) and the brake cam arm (8682 for front wheel and 8380 for the rear wheel) from the brake cam (8371).

- Take the brake cam out of the brake plate.

- Where applicable (brake plates 9267 for the front wheel and 9546 for the rear wheel) remove the 18mm diameter brake shoe pivot (9287).

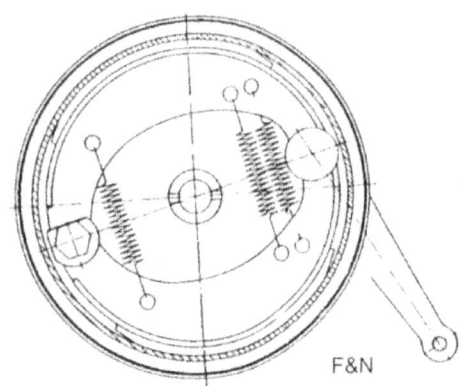

Assembling is done in the reverse order.

Fit two return springs at the cam end of the shoes, one behind and one in front of the shoes, either in the two closely adjacent drilled holes (in fabricated steel shoes) or in the same large hole (in the cast shoes with pivot eyes)

Maintenance and repair of brakes

Brake drums

The 150mm brake drums are riveted to the front and rear hubs. The front wheel with a 180 mm brake drum has an integrated brake drum. Only the 180 mm rear wheel brake drums can be separated from the hub. This means that a workshop must be found which has the equipment to machine non-removable brake drums by mounting the complete wheel in a lathe. The alternative is to dismantle the wheel, and for the 150mm brakes, this requires drilling out the rivets which secure the drum to the hub. The need for machining the brake drum depends in part on how evenly it is worn, and whether ovality is present. Ovality is caused by faulty spoking and is not resolved by machining the drum.

The number of times a brake drum can be re-machined is limited -

more limited in the case of the 150mm drums than the 180mm. Remember that re-machined brake drums need oversized brake shoes. For the 180mm brake, 1st oversize is 181 mm and 2nd oversize 182 mm.

The 150 mm brake drums and 180 mm front drums can be so worn that the only option is either to buy new parts or have the old ones built up by welding and then machined to the standard dimension.

Brake shoes and linings

The brake linings can be renewed. Riveted linings in particular must be renewed before the rivet heads damage the brake drum. Re-machined brake drums are oversize and must be fitted with thicker linings.

Brake linings are commercially available for 180mm drums which have been machined out to 181 mm or 182 mm. When renewing the brake linings, they should preferably be bonded to the shoes, even if supplied pre-drilled for riveting. The leading edge of all brake linings should be chamfered.

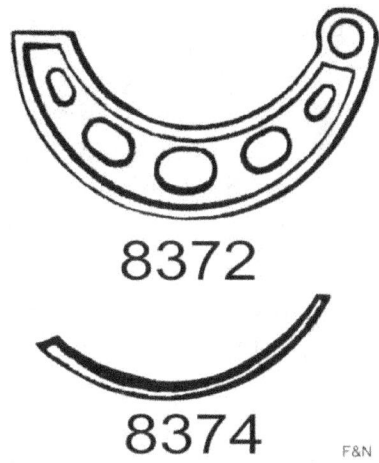

If 180 mm brake linings are worn but serviceable, and brake arm travel is excessive, a first step could be to remove the brake arm (8682 front, 8380 rear) and turn it over so that the "S" stamping is facing outward.

Brake cam pivot wear

In worn brake plates, particularly aluminium ones (9267 and 9546), the brake cam pivot may show excessive play. If that is the case, the pivot hole can be drilled out and a bush with inner diameter 18mm pressed into place.

Reduced braking effectiveness
If the brakes have less than standard effectiveness, despite brake drum machining and replacement of the linings, check the drums and linings for contamination with grease, oil or water. Especially with the 150mm brakes, there is unfortunately a risk that grease from a wheel bearing will enter the drum.

Check also on whether the linings are fully contacting the drum. If there are high spots file them down, reassemble the brake, and test for better contact. Repeat this procedure aiming to get as much of the lining contacting the drum as possible.
Take care: brake linings may contain asbestos. Always use respiratory protection when working, and vacuum all dust for safe disposal.

Adjusting the brakes

Adjusting the front brake

- Check the free movement of the handlebar brake lever. This should be no more than ¼ of the total travel. Adjustment is by means of

the brake cable adjusting sleeve nut (7208).

- Check the brake's effectiveness. Roll the motorcycle forward and operate the brake. The front brake must lock, causing the energy of motion to be absorbed by the forks.

- Check the angle between the brake arm and the brake cable at the brake plate (8682). If this angle is more than 90° with the brake applied, remove the brake arm and reverse it so that the "S" stamping is visible. If this has already been done, the brake should be dismantled for lining replacement and possibly drum re-machining.

Adjusting the rear brake

- Check the travel of the brake pedal: this should be 15 mm. Adjust it by means of the sleeve nut (7256) on the brake pull rod (7670).
- Check the brake's effectiveness. Ride the motorcycle at moderate speed on a straight dry and preferably traffic-free road and apply the brake firmly. The rear wheel should lock. *Be careful* not to lose control of the motorcycle during testing.

- Check the angle between the brake arm (8380) and the brake pull rod (7670). If this angle is more than 90° with the brake applied, remove the brake arm and reverse it so that the "S" stamping is visible. If this has already been done, the brake should be dismantled for lining replacement and possibly drum re-machining.

Adjusting the sidecar brake

First, adjust the motorcycle rear wheel brake (see above).

- Check the effectiveness of the sidecar brake. Ride the motor cycle, with sidecar, at moderate speed on a straight dry and preferably traffic-free road and apply the brake firmly. The outfit should not pull to either side. If the sidecar outfit pulls to the right, the sidecar brake adjuster (9598) on the sidecar brake pull rod (8499) must be slackened. If it is pulled to the left, the sidecar brake adjuster must be tightened.

Side car brake rod

Check the angle between the sidecar brake arm (8476) and the sidecar brake pull rod (8499). If this angle is more than 90° with the brake applied, remove the brake arm and reverse it so that the "S" stamping is visible. If this has already been done, the brake should be dismantled for lining replacement and possibly drum re-machining.

Frame

The Nimbus-C main frame members are 40 x 8 mm C 35 flat steel. If a frame has been twisted or bent out of alignment, it can - within limits - be straightened. Frame straightening must always be done cold. Frames which have been subject to heat as a result of fire are unsafe and repair should not be attempted.

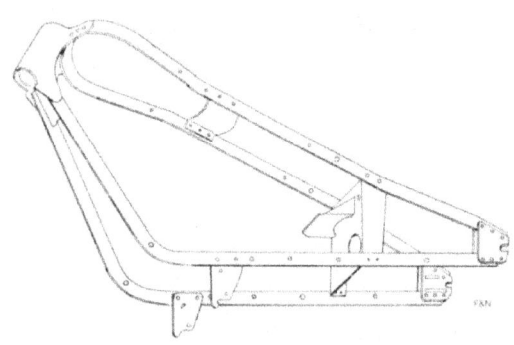

Frame straightening

Misalignment of a frame can be checked by placing a straightedge along each of the two lower frame members. A rod fitted accurately through the centre of the steering head must project exactly midway between the two straightedges. If this is not so, the frame must be straightened in a workshop that has a frame alignment bench.

The rear fork of the frame can be checked with the rear fork gauge (9030). With the gauge fitted into the frame, a 0.2mm feeler should fit between the frame (without paint) and the buttons on the gauge. If the test fails, the rear fork must be realigned with the realignment tool 9034.

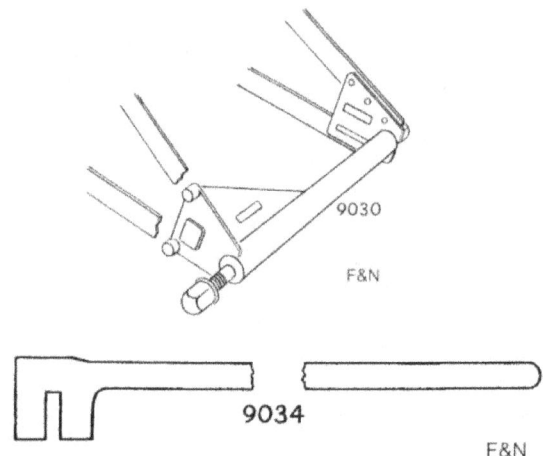

In most cases the rivets have to be drilled out and frame parts have to be removed in order to straighten the frame. New rivets must be fitted with the use of a dedicated rivet setting tool. Riveting of a Nimbus-C frame must always be done *cold*. Welding repair to the few places where some frames were originally welded, such as the steering head, is allowable, but is a job for specialists.

Note that other than this, no welding should be done on any load-

bearing component: frame, forks or handlebar.

Frame modifications

While each version of the Nimbus-C frame (7310, 8244 or 8244-2) precisely met the designed requirements at the time of manufacture, it is possible make changes to meet different, later needs.

In the 1960's the factory supplied 'universal frames', ready to accept any part manufactured between 1934 and 1959. If such a frame is not available, the frame to hand can be modified in a number of ways.

The steel used in the frame is difficult to drill. Holes must be started with a bit of a smaller diameter than the finished hole. Drill bits must be cooled with cutting oil or soapy water.

Damaged threads in the frame can be drilled out and repaired with a thread insert.

Changing the seat suspension

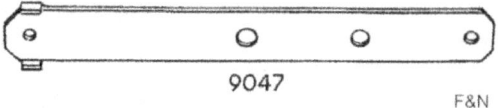

To drill the holes needed to convert from coil springs to rubber suspension (or vice-versa) drilling template 9047 is used. If this template is not available, the holes can be drilled at both sides of the frame by referring to the diagram. Finish drilling with an 8.5mm bit.

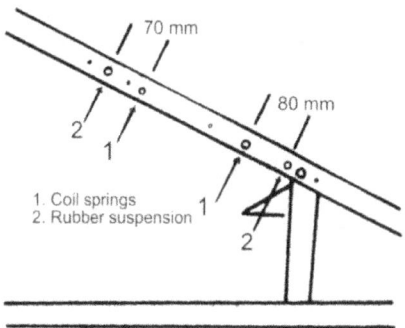

Lever pivot for early pattern clutch release arm and lever

To fit the pivot (8416) for the lever (8415) which actuates the clutch release arm (7076) of the early pattern gearbox, drill a 7.0mm hole at the position indicated on the diagram. Cut an 8M x 1.0 thread.

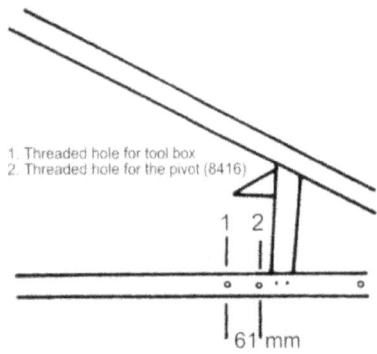

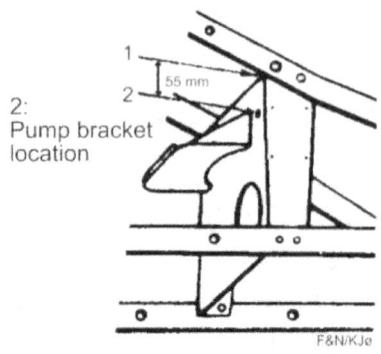

Pump bracket location

To adapt the frame to accept the rearmost pump bracket behind the saddle, drill a 7.0mm hole as indicated on the diagram and tap an M8 x 1.0 thread.

To fit the pump bracket under the saddle, drill a 6.5mm hole as indicated.

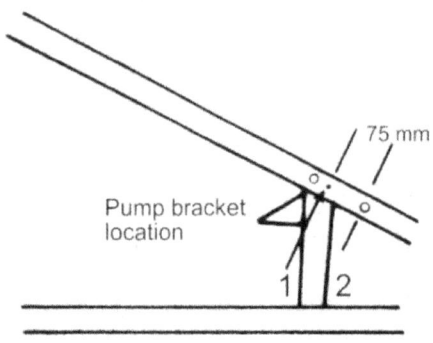

Changing the drive-shaft and frame cross-member

Modifying a later frame to suit an early pattern rigid drive shaft (7131) involves removing the frame cross-member (7303-2) with the 69mm hole and replacing it with a cross-member (7303) with the smaller 42mm hole. The rivets can be drilled out. The frame cross-member (7303-2) has the battery carrier (7478-2) attached to it, either riveted (up to frame number 7500) or welded.

Fitting a military prop stand

Modifying the frame to fit a military prop stand requires that a 12mm hole has to be drilled in both upper frame members as the stand is fitted to a stay which braces the frame. The holes must be drilled 50 mm in front of the regulator bracket mounting holes.

Steering lock

From 1956, frames were provided with a steering lock. A frame lacking this lock could be modified to accommodate one. Instructions were given in a technical circular (*Circular 82, 03/05/1956*). However the required lock cylinder and associated parts are no longer available. Keys for the steering lock are numbered. The numbers are listed in the stock records of 1956 and later. Key numbers should be recorded in case a replacement is needed.

Maintenance of the steering lock

The steering lock and key have to be regularly lubricated with special lock oil, especially during freezing weather. Never leave the key in the lock. It may be bent or broken by the left fork leg.

Knee pads

Knee pads (7405) and inserts (7654) are fitted to brackets (7653) fixed to the upper frame member by M6 x 12mm bolts. To suit rider preference, the brackets can be fitted to the frame by any of the four holes in the brackets. Use both flat and spring washers, and to avoid damaging the petrol tank do not use bolts longer than 12mm.

Footrests and footrest rubbers

Foot rests (7357) are designed such, that they can be fitted to accommodate the rider's leg length (within certain limits). The brake pedal pivots on the shoulder of the right footrest. Wear between the brake pedal and the footrest can be partly corrected by fitting one of the other footrests at this point. Footrest rubbers manufactured according to the original drawings will not easily rotate on the footrest and use of some silicone oil or detergent will help fit them.

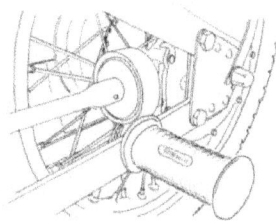

Tool box

Empty the tool box before removal to prevent the brackets from being bent. When fitting, the lid should be on the right side, otherwise kickstart action may damage the hinge or dislodge the lid. In the military version, the tool box is mounted behind the pillion seat, attached to the frame by the bolts that secure the right and left brackets (9680/9681) for the seat suspension rubbers.

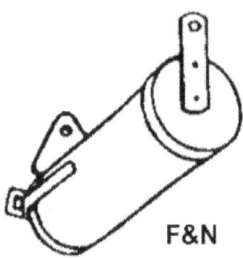

F&N

Centre stand

Removing the centre stand

- Take the motor cycle off the centre stand and have an assistant steady it.

- Remove the two centre stand springs (7585) from the stand and the frame brackets, using a screw driver as a lever.

- Undo the nuts of the two stand pivot bolts (7273 or 8329), remove the pivot bolts and draw the stand back and down.

Fitting the centre stand

- Locate the centre stand (7300 or 8328) between the centre stand frame brackets with the cross-brace on the underside.

- Fit the pivot bolts (7273 or 8329) from the inside so that the spring washers and nuts (7260) will be outside the frame brackets. Fit the washers and nuts but do not tighten them until the springs have been fitted.

- Fit the centre stand springs. This can be done in several ways. One way is to locate the spring on the stand and then to pull the other end of the spring up to the frame attachment point using a cord or wire passed up between the engine and the frame.

- Put the motorcycle on its centre stand and tighten the centre stand pivot nuts.

Petrol tank

Removing the petrol tank

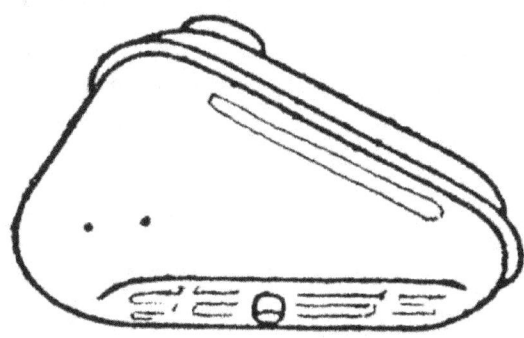

- Close the petrol tap and remove the fuel line (7683, 8183 or 8577).

- Remove the frame cover plate (7852, 8301, 8417 or 10512) or the clamping bracket (7451 or 7451-2) at the rear of the tank.

- At the front of the tank, remove the single clamping bracket (7451 or 7451-2) or the left (10922) and right clamping brackets (10923).

- Remove the metal tank badges (10664) if present.

- Lift the petrol tank carefully out of the frame.

- Remove the tank cap (7455).

- Empty the tank, remove the petrol tap (7663) and remove the rubber seating ring (7657).

Fitting the petrol tank

Fitting is done in the reverse order (but wait until the tank is fitted before refilling it).

The rubber seating ring (7657) will fit more readily if it is lubricated with silicone spray or soapy water.

Repairing the tank

Leaking tank cap:
Remove the plate (7457) and spring plate (7458) and renew the gasket (7486). If the problem is not solved, bend the spring plate flanges slightly *up* in order to provide a firmer grip at the filler neck.

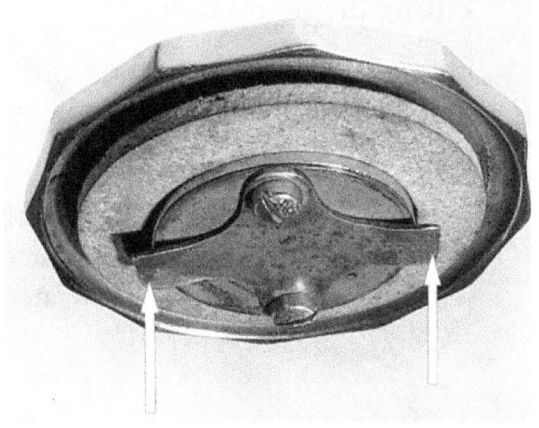

Leaking petrol tap:

The most likely cause of a leaking petrol tap is a dried-out cork gasket. The tap must be dismantled and depending on the tap action (turn or slide), the the cork gasket or gaskets can be replaced.

Leaking petrol tank:

If the leak is in the bottom of the tank, a repair may not prove successful in the long term, as engine heat causes the tank to flex, which can lead to cracking around the repair.

Take care: Any open flame or spark associated with soldering or welding, even with an emptied tank, can lead to explosion of petrol vapour. To reduce this risk, the tank must be filled with water and thoroughly flushed to rinse out fuel residue before any welding repair is attempted.

If the leak is at the seam between the upper and lower parts of the tank, the leak can be repaired by using a suitable two-component resin.

Telescopic forks

There are six versions of telescopic forks to be found on Nimbus-C motorcycles:

The four 'early' internal slider versions have <u>internal</u> sliders, and one of these versions is oil-damped. (Danish owners generally refer to these as 'low' forks).

The two 'later' versions have <u>external</u> sliders and are oil-damped. (They are known in Denmark as 'high' forks).

Removing the complete fork assembly

Put a 20-25mm board under the centre stand and pull the motorcycle up onto the stand.

- Remove the front wheel and the front mudguard (see the applicable section).

- Remove the handlebar (see the applicable section). To access the fork stem lock nut (2984) and threaded bearing cone (3111) it will be useful to remove the head light, and for 'later' external slider fork machines the horn as well.

- Remove the lock nut (2984) from the handlebar. Use the special tool

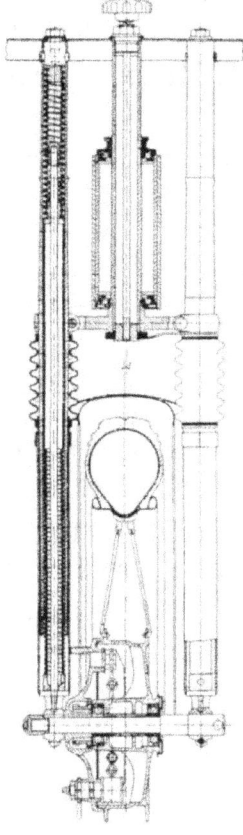

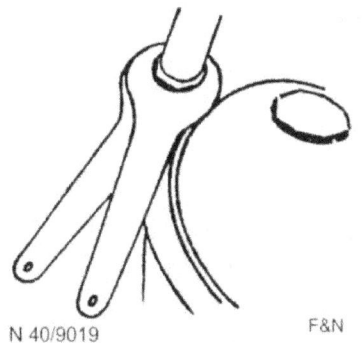

N 40/9019   F&N

N40/9019 or a 46mm open ended spanner and remove the internal tab washer (8245) fitted to machines from frame number 2551.

- Remove the dust cover (2988), which can be prised off with a screwdriver.

- Slacken and remove the threaded bearing cone (3111) while supporting the forks. Spread a cloth on the floor and collect all 24 balls - a magnet will be useful.

- Withdraw the fork stem assembly from the steering head.

Fitting the forks

Because of the weight of the forks it is advantageous to first mount only the fork stem in the steering head, and then to fit all other fork components to it.

Preparation, assembly and fitting

Prepare the forks for mounting in the steering head.

- Thoroughly clean the thread of the fork stem and test the fit of the threaded bearing cone (3111) by screwing it all the way down the stem.

- For the 'early' internal slider forks: Fit the washer (3590), felt ring (2986), and the plain (non-threaded) bearing cone with inner diameter of 35.08 mm (2989) on the fork stem

- For 'later' external slider forks: Fit the dust cap (9474) and the plain (non-threaded) bearing cone with inner diameter 36.0mm (9599) on the stem. *Take care:* There are two different sizes of plain cones. Make sure the correct one is used.

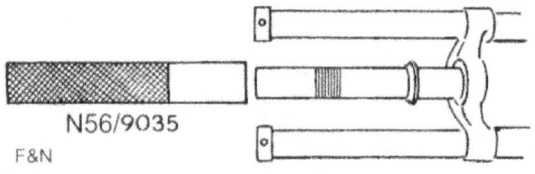

- Prepare the steering head in the frame for fitting the forks. Fit the two bearing cups (2991) in the steering head. Drive both home in the steering head bearing seats using special tool N45/9024.

- Apply grease to the plain bearing cone (2989 or 9599) at the bottom of the fork stem. Place all 24 ¼' balls (3846) evenly spaced around the cone.

- Apply grease to the upper bearing cup (2991) of the steering head. Place 24 ¼' balls (3846) evenly spaced around the cup and put the threaded bearing cone (3111) in place.

- Carefully insert the fork stem from below into the steering head and through the upper bearing cup. Support the forks while threading on the upper bearing cone, first by hand, then with special tool N40/9019 or a 46 mm open ended spanner.

- Press the dust cover (2988) in place over the upper bearing.

- Slide the internal tab washer (8245) down to the threaded cone. Early machines (frame numbers below 2550) were not supplied with this tab washer.

- Fit the lock nut (2984) to the fork stem.

- Adjust the threaded bearing cone (3111) so that all slack in the steering head is eliminated, but the forks still move freely. Tighten the lock nut and check the adjustment again. Use special tool N40/9019) or 46mm open ended spanners. If the forks have been fitted to the steering head without the sliders and springs these must be installed now, before fitting the handlebar, head light, and horn.

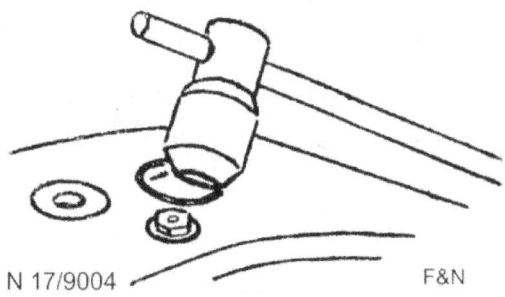

N 17/9004   F&N

- Fit the handlebar (see the applicable section). Fit the throttle and clutch cables if they have been when removed, and for 'early' internal slider forks, fit the horn.

- Fit the front mudguard and the front wheel (see the applicable section).

Dismantling the forks

Most maintenance and repair procedures can be done with the forks in place. Dismantling is easier to do in this way because of the weight of the whole fork assembly.

Dismantling inner slider forks without oil damping

- Put a 20-25mm board under the centre stand and pull the motorcycle up onto the stand.

- Remove the handlebar (see the applicable section).

- Remove the gaiter clips (7659 or 8534) and the leather gaiters (7658).

- Remove from the 1938 pattern forks (with friction dampers) 2 caps (8440) with grease nipples (7649), damper springs (8441) and damper pistons (8439).

- Remove the guide screw (7184) from each stanchion.

- Remove the cross-pin (7183) from the threaded plug (7175) on each stanchion

- Pull the fork sliders (7627) with springs (7313 and 7878) and threaded plug (7175) out of the stanchions.

- Remove the threaded plugs (7175) from the springs. If the plugs are difficult to remove, clamp the spring in a vice and unscrew the plugs using a screwdriver as a tommy bar.

- Take the springs out of the fork sliders. The inner spring (7878) will pull straight out (7313). The main spring is fitted onto a threaded pin at the bottom of the inner member and has to be unscrewed.

- Remove the fork stem assembly (see above).

- Drive the unthreaded bearing cone (2989), washer (3590) and felt seal (2986) out of the steering head. Use a punch and hammer.

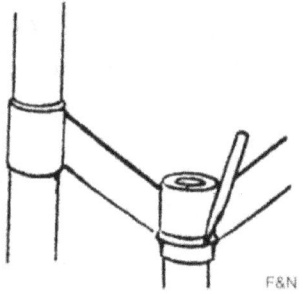

Note: The stanchions of the 'early' internal slider forks may have been reconditioned by having 63mm long brass bushes (9461) fitted. If such a bush has to be removed, a ¾' bottoming tap can be screwed partly into it (which will make the bush unusable) and the tap and bush driven out using a long rod as a drift, working from the other end of the stanchion.

Dismantling inner slider forks **with** oil damping

- Put a 20-25mm board under the centre stand and pull the motorcycle up onto the stand.

- Remove the handlebar (see the applicable section).

- Remove the gaiter clips (8717 and 8703) or cut the gaiter retaining wire (9425) and remove the rubber gaiters (8702).

- Remove the guide screw (7184) from each stanchion.

- Remove the cross-pin (7183) from the threaded plug (8688) on each stanchion

- Pull the fork sliders (8695 and 8696) out of the stanchions complete with spring assemblies.

- Drain the oil from the reservoirs.

- Remove the upper threaded plugs. If necessary, clamp the spring (8690) in a vice and unscrew the plug using a screwdriver as a tommy bar.

- Remove the upper spring from the lower threaded plug (8687).

- Remove the special sleeve nut (8705) from the slider pullrod. Use a pair of adjustable pliers or similar tool, while pulling the centre and lower springs (8691 and 8689) down.

- Remove the threaded plug and centre and lower springs from the pullrod. *Note:* The pullrod *cannot be removed.*

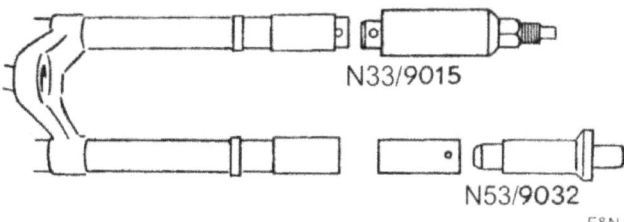

- If the stanchion bushes (8692) must be renewed, they can be removed by using special tool N33/9015). Heat the stanchions-if necessary.

Note: Stanchions may have been modified and longer 63 mm bushes (9461) may have been fitted. If such a bush has to be removed, or if the original bush cannot be removed by using a puller, then a ¾' bottoming tap can be screwed partly into it (which will make the bush unusable) and the tap and bush driven out using a long rod as a drift, working from the other end of the stanchion.

Dismantling external slider forks

- Put a 20-25mm board under the centre stand and pull the motorcycle up onto the stand.

- Remove the handlebar (see the applicable section).

- Cut the gaiter retaining wire (9425) and remove the rubber gaiters (9314).

- Remove the cross-pin (7183) from the threaded plug (9305) on each stanchion.

- Pull the sliders (9284 and 9285) off the stanchions complete with spring assemblies.

- Pour out the oil from the sliders.

- Remove the upper threaded plugs. If necessary, clamp the spring (8690) in a vice and unscrew the plug using a screwdriver as a tommy bar.

- Remove the upper spring from the lower threaded plug (8687).

- Remove the special sleeve nut (8705) from the slider pullrod:

* For 1948 pattern fork legs with three compression springs and a distance piece (9547) below, use a pair of adjustable pliers or similar tool, while pressing down the centre and lower springs (8691 and 8689).

* For 1950 pattern fork legs with a rebound spring, 2 compression springs and a central distance piece (9589), use a pair of adjustable pliers or similar tool, while pressing down the rebound spring (9333), threaded plug (8687), distance piece (9589) and lower spring (9588).

- Remove the threaded plug, springs and distance piece from the pullrod of each slider.

- Remove the pullrod (9306), if necessary by clamping it in a vice and unscrewing the slider.

- Invert the slider allowing the oil restrictor (9590) to drop out.

Dismantling the external slider fork stem assembly

- Remove the two yoke clamping nuts (7260 or 9929).

- Drive the clamping bolts (9301) out of the yoke.

- Ease the yoke clamp slightly open, using a flat blade screwdriver as a wedge, and remove the stanchions. (9282).

- If the fork bushes (9309 and 9310) are to be renewed, the stanchion must be clamped in a vice and the bushes removed with a sharp chisel.

Note: If an upper bush (9310) appears to be missing, it is probably stuck in the slider. If it cannot be removed after heating the slider, a puller must be made. A 34mm disc is needed, with a central 6mm threaded hole. It will need to have two opposite flat edges cut, to allow it to be dropped through the bush to the bottom of the slider. A long 6mm threaded rod can then be screwed into the disc, to pull it out together with the bush.

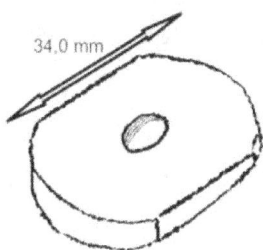

Assembling the forks

Assembly of the dismantled forks can be completed with the stem assembly fitted to the frame. The motorcycle will need to be on its centre stand with a 20-25mm board beneath it.

Assembling inner slider forks **without** oil damping

Normally, these forks are not fitted with bushes. If the stanchions have worn out of round, they may have been bored out to 30mm and fitted with 63mm long bushes (9461). Use special tool N53/9032 if they need to be removed.

- Thoroughly grease the springs (7313 and 7878) and place the main springs in the sliders (7627), screwing each to the threaded plug at the bottom of the slider. Insert the auxiliary inner springs in the main spring.

- Fit the threaded plugs (7175) to the main spring.

- Thoroughly grease the sliders and insert the sliders, springs and threaded plug up into place in the stanchions.

- Fit a cross-pin (7183) through the threaded plug in each stanchion.

- Fit grease nipples (7649) to the stanchions. Each stanchion has three threaded holes and the grease nipples should be fitted in the upper and in the lower holes.

- For 1938 pattern forks with friction damping, fit the damper pistons (8439), springs (8441) and covers (8440) with grease nipples (7649).

- Fit the guide screw (7184) to each stanchion in the threaded hole just below the upper grease nipple. It is essential that the guide screws engage the longitudinal slot in the sliders. Turn the sliders as needed in order to bring this about.

- Fit gaiters (7658) to the fork legs and secure each of them with the gaiter clip (7659 or 8534).

Assembling inner slider forks **with** oil-damping

Assembly of the dismantled forks can be completed with the stem assembly fitted to the frame. The motorcycle will need to be on its centre stand with a 20 - 25mm board beneath it.

- Fit a 75 mm long brass bush (8692) in each stanchion. Use special tool N53/9032. If the forks are to be modified for a front wheel with 150mm brake the standard 75mm bushes must be replaced by 63mm long bushes (9461) and matching sliders (9460) must be fitted.

- Fit the longest spring (8689) over the pullrod in each of the sliders (8695 and 8696) followed by the smaller (19.5mm) diameter spring (8691).

- Fit the threaded plug (8687) and the special nut (8705) to the slider pullrod. Use a pair of adjustable pliers or similar tool, while pulling the centre and lower springs (8691 and 8689) down.

- Fit the top spring (8690) onto the threaded plug (8667) and fit the upper threaded plug (8688) to this spring.

- Apply some grease to the upper part of the springs and insert the fully assembled sliders in the stanchion's assembly. The slider with the spindle clamp (8696) is fitted to the left and the slider with the plain hole (8695) is fitted to the right.

- Fit a cross-pin (7183) through the threaded plug in each stanchion.

- Fit the guide screw (7184) to the threaded hole in each stanchion. It is essential that the guide screws engage the longitudinal slot in the sliders. Turn the sliders as needed in order to bring this about.

- Fill the slider oil reservoirs up to the oil level hole. For sliders (9460) used with a 150mm brake the level holes are on the outer side of the oil reservoirs. Fit the oil level plug (8725 or 7749) with fibre washer (8724).

- Fit the gaiters (8702) with clips (8717 and 8703) or wire (9425).

Assembling external slider forks

- Fit the fork bushes to the stanchions: first the upper bush (9310) and then the lower bush (9309). Heat the bushes and drive them into place, if necessary, with special tool N56/9035. *Make sure* that any burrs in the bushes are removed before fitting. If the upper bushes are a loose fit after cooling due to wear on the stanchion, the bushes need to be secured by high-grade Loctite or tinning. *Make sure* that any excess solder is removed.

The bush should be finished to a diameter of 34mm.

- Fit the stanchions (9282) into the fork stem assembly (9278). Turn the stanchions to align the ground recesses for the clamping bolts (9301) with the bolt holes. There is a special gauge set (Tools N36A, B/9017, and 9018) to assist correct assembly of the stanchions and fork stem. Tighten the yoke clamping nuts (7260) or (9929). *There is a risk of overtightening these nuts.* Check by eye to verify if alignment is correct. Further assembly can best proceed after the fork stem assembly and stanchions have been fitted into the steering head. (See section on fitting the forks)

- Fit the pullrods (9306) to the fork sliders. If necessary, clamp the pullrod in a vice and screw the slider onto it

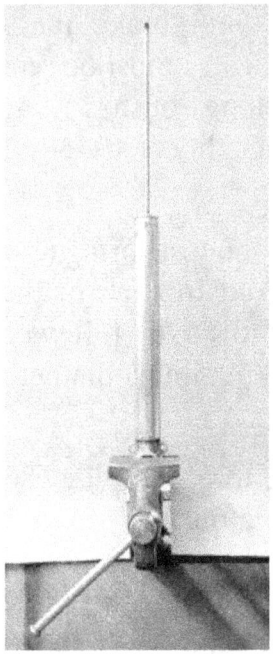

- Complete the assembly of the slider as follows.

For 1948 pattern forks, for each slider:

* Slide the distance piece (9547) and the 241mm lower spring (8689) over the pullrod to the bottom of the slider.

* Fit the middle 189mm spring (8691) and threaded plug (8687) above the lower spring and compress the springs sufficiently to allow fitting of the sleeve nut (8705) to the pullrod.

For 1950 pattern forks, for each slider:

* First slide the oil restrictor (9590) over the pullrod with its larger opening uppermost, then slide the lower spring (9588 or 9587) in place.

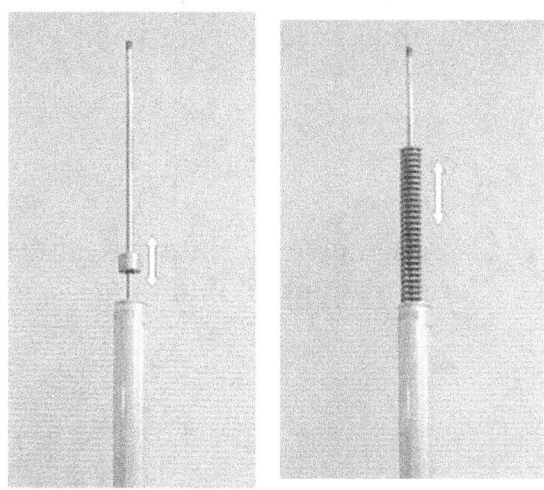

* Add the distance piece (9589), threaded plug (8687) and rebound spring (9333) to the pullrod. Compress the springs sufficiently to allow fitting of the sleeve nut to the pullrod.

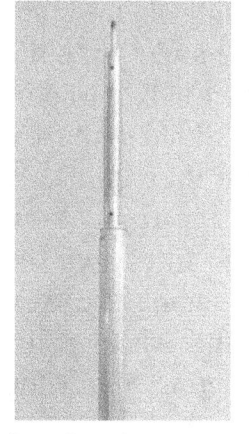 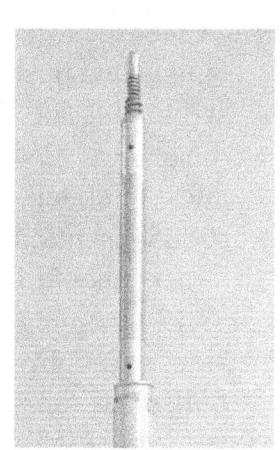

* Fit the upper 185mm spring (8690) to the threaded plug and grease the spring.

- Fit the upper threaded plugs (9305).

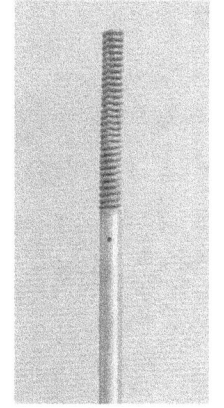 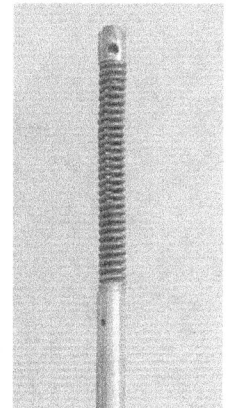

- Fit the rubber gaiters (9314) loosely to the stanchions.

- Enter the fork springs in the stanchions and push the slider onto the stanchion. The slider with the spindle clamp (9284) is fitted to the left and the slider with the plain hole (9285) is fitted to the right. Both sliders should be so fitted that the 'OIL' stamping is facing the outside, away from the wheel. This will place the clamping nut at the rear of the left fork leg

- Fit a cross-pin (7183) through the threaded plug in each stanchion.

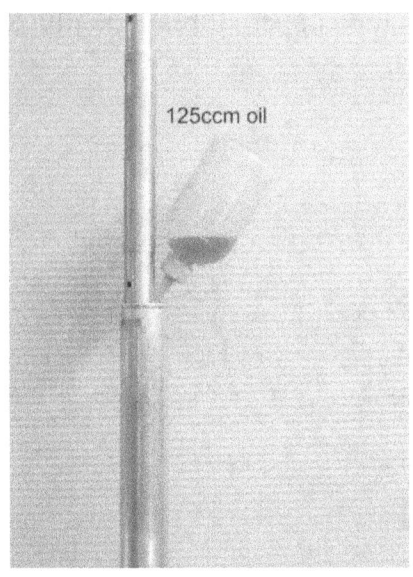

- Pour oil into the sliders.

NB: the 'OIL' level mark indicates the oil level in an *empty* slider. The correct amount of oil to be supplied is about 125ml which will fill a slider to within 150mm from the top when the slider contains all working parts. Use SAE 30 mineral oil, without additives, to prolong the life of the rubber gaiter.

- Fit the rubber gaiters (9314) to the sliders and secure them with wire (9425).

Fork maintenance

Maintenance of internal slider forks <u>without</u> oil damping

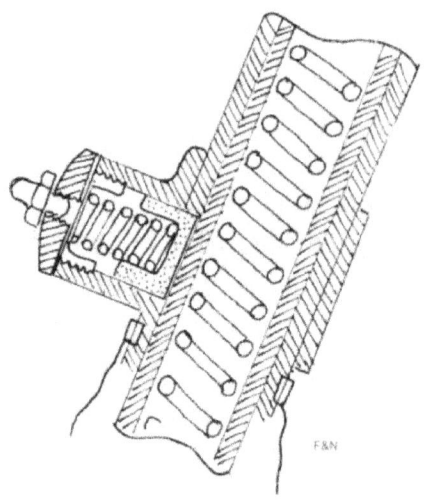

The internal slider forks, with or without friction dampers, are grease-lubricated. Grease nipples

(7649) are fitted in the stanchions and in the friction damper covers (8440).

Greasing must be done every 1000 to 2000km.

Maintenance of internal slider forks <u>with</u> oil damping

Internal slider forks with oil damping are partly oil-lubricated. It is good practice to grease the upper parts when assembling. Oil in the reservoirs should be kept up to the oil level screw at the front or side of the reservoir. Use SAE 30 mineral oil, without additives, to prolong the life of the rubber gaiter. Check the oil level every 5000km and change the oil when overhauling the forks or in the event of gaiter damage.

Maintenance of external slider forks

External slider forks are oil-lubricated. It is good practice however to grease the upper parts when assembling. The oil level in the sliders should be within 150 mm from the top. There should be only oil and no water in the slider. With already assembled sliders ignore the 'OIL' mark, which only indicates the oil level for an uninstalled empty slider. Top-up the sliders when required with SAE 30 mineral oil, without additives, to prolong the life of the rubber gaiter. Check the oil level every 5000km and change the oil when overhauling the forks or in the event of gaiter damage.

Fork repair

Fork leg realignment

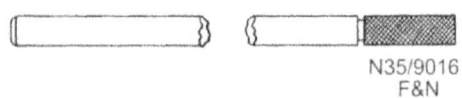
N35/9016
F&N

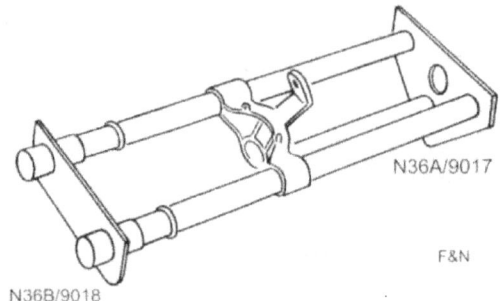
N36A/9017
N36B/9018
F&N

Realignment of the fork legs must be done cold. A close-fitting pipe is used, which is slid over one end of the leg to be aligned, while the other end is fixed in a stout vice. During the process, alignment should be checked regularly by use of special tool N35/9016 or a straight piece of round bar with a diameter of 25.2mm. Finally, alignment is visually confirmed by using special two-plate gauge set N36-A-B/9017-18: when the edges of the gauge plates are seen to lie in the same plane, alignment is correct. The gauge plates are suitable for all fork versions. Alignment of external slider forks must be done with the forks removed from the frame (see the applicable section).

Trail should be measured after fork alignment. The trail should be 60mm for internal slider forks and 65mm for external slider forks. Note that wheelbase (front to rear spindle distance) differs according to fork type, the figures being 1410mm for internal slider forks and 1435mm for external slider forks.

Fork wear

Fork life can be prolonged by regular and correct maintenance. For forks fitted with friction damping, the sliders can be exchanged, left for right, which means that the damper piston will be working on a different area of each slider.

Both stanchions and sliders (with or without oil reservoirs) can be renewed, often on an exchange basis, so do not scrap any parts.

Wear on fork springs reduces their effectiveness. Rusty springs and springs that clearly show marks of wear, should be renewed. Worn bushes should always be replaced.

Damaged forks

A noisy or rattling external slider fork is often an indication that one or both upper bushes have become loose on the stanchion, causing too much play. In that case it may be necessary to renew both stanchions and bushes.

If an oil damped fork is bottoming, this may be an indication of insufficient or water-contaminated oil, or worn out springs or bushes. Grease-lubricated forks are also subject to bottoming.

A cracked fork yoke is highly dangerous, as is a stripped fork stem thread, and either of these conditions must be corrected before the motorcycle is ridden.

A split or torn rubber gaiter allows water contamination of the oil, leading to poor damping and lubrication and to rust damage.

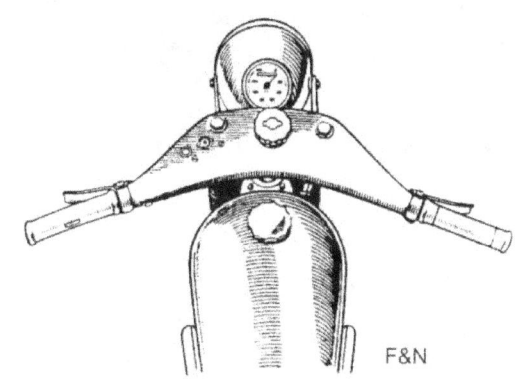

Handlebar

There are two versions of handlebar used on Nimbus-C motorcycles. The 'early' version was fitted to machines with internal slider forks, and the 'later' to machines with external slider forks.

Removing the handlebar

- Remove the steering damper knob (7648, 7648-2 or 8897) and double spring washer (3864) if fitted. This will allow the steering damper plate and nut (7647 or 9308) and friction disc (7177) to be removed from below the fork yoke.

- Remove the fork stem nut (7172) with special tool N17/9004 or a 27mm spanner (for frame numbers 1301 to 2600 use a 19 mm spanner) and remove the stem nut washer (7576 or 9302).

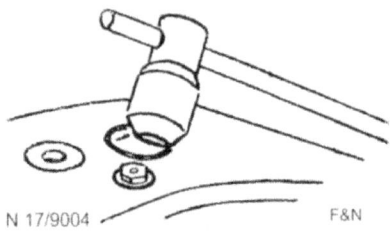

- On the early handlebar remove the two stanchion nuts (7260) and washers (7527). On the later handlebar remove the two stanchion securing bolts (9304) and washers (9303).

- Remove the throttle cable (7673, 8585 or 8927) from the hooked cross-pin (7211) of the twistgrip.

- Remove the clutch cable (7672, 8413 or 8944) and front brake cable (7671, 8248, 8707 or 8945). Remove, if fitted, the circlips from the clutch and brake lever adjustment sleeve nuts (7208) and take them out of the lever rollers (7207)

- Remove the ignition switch (7345 or 8880) by releasing the two fixing screws (7834) and withdrawing the switch and connected leads from the lighting twistgrip.

- On the early handlebar take out the two securing bolts (7265) of the horn (7384) and its leads. Remove the horn (7384). On the later handlebar disconnect the leads from the handlebar to the horn.

- Lift the handlebar from the front fork assembly.

Dismantling the handlebar

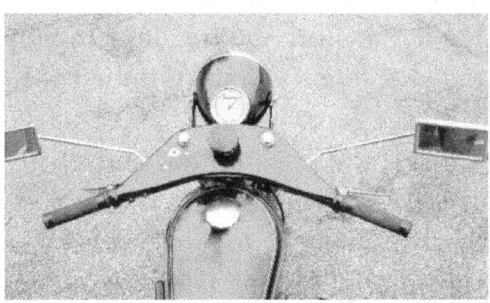

- Remove both lever pivot screws (7204 or 7204-2) from the clutch lever and brake lever (7198) and prise the levers off the threaded bosses on the handlebar.

- On the early handlebar remove the speedometer (7386 or 8366) and instrument lighting (4496) if fitted, the ammeter (7665) or the ignition warning light (8290) with its identification plate (8287) and backplate (8288).

- Remove the lighting twistgrip (7692 or 8893):
* Remove the threaded plug (7650) securing the spring (7074) and ball (3845) for the lighting switch and take out the spring and ball.
* Pull out the cross-pin (7203 or 8870), the spring (7460) and, if present, the steel (8789) and fibre washers (8790).
* Take the lighting twistgrip out of the handlebar.
* Remove the rubber handgrip (7205, 7205-2 or 7205-3).

- Remove the throttle twistgrip (7691 or 8899):
* Remove the throttle twistgrip stop screw (7488).
* Pull out the hooked cross-pin (7211).
* Remove the spring (7460) and for the later handlebar, the steel (8789) and fibre washers (8790) and the distance tube (8888).
* Take the throttle twistgrip from the handlebar. *Be careful:* the later handlebar is fitted with a bearing including 28 uncaged balls (5608) which will be released by this action. Set them aside for reassembly.
* Remove the rubber twistgrip (7205, 7205-2 or 7205-3).

- On the early handlebar, remove the horn button. The cover may be screwed on or held by a spring clip. Take out the button and spring and remove the two screws (5509) in the base of the switch. On the later handlebar remove the horn button by taking out the two mounting screws (5509).

 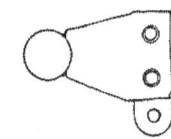

- On the later handlebar, remove the ignition warning light lens (9273), which is held in place by crimping of the lens rim on the underside of the handlebar. Bend back the crimping as needed to release the lens.

- Remove the ignition keyhole trim (7461 or 7461-2) by driving out the two rivets (5435), from the underside of the handlebar.

Assembling the handlebar

- Fit the identification plate (7461 or 7461-2). Use two rivets, (5435), or supplement or replace the rivets with a modern type of adhesive.

- On the early handlebar, fit the identification plate (8287) and backing plate (8288) with two screws (8292), lock washers (8293) and nuts (5040).

- On the early handlebar, fit the horn push button (7385):
* Fit the horn button housing by the two screws (5509) in the base.
* Fit the horn lead.
* Insert the push button and spring and fit the switch cover which either screws on or is held by a spring clip. On the later handlebar, fit the horn button (9324) with two screws (5509).

- Fit a rubber handgrip (7205, 7205-2 or 7205-3) to both twistgrips. *Note:* This can wait, until the twistgrips have been fitted to the handlebar. When fitting new rubber handgrips use a lubricant, for example a detergent, which will not damage the rubber.

- On the later handlebar, fit the ignition warning light lens (9273). Press the lens into the handlebar and crimp the underside sufficiently to secure it Supplement or replace the rivets with a modern type of adhesive.

- Fit the lighting twistgrip (7692 or 8893):
* Locate the twistgrip in the left of the handlebar

* Fit the spring (7460) and cross-pin (7203 or 8870) and if applicable, the steel (8789) and fibre washers (8790)

* Grease and insert the ball (3845), spring (7074) and fit the threaded plug which retains them.

- Fit the lighting switch (7345 or 8880). Turn the lighting twistgrip fully "open" (clockwise) and turn the switch rotor to align with the cross-pin in the twistgrip. Slide the switch into place and secure the switch with the two screws (7834). Make sure that the ignition switch (7336) clears the ignition key trim (7461 or 7461-2).

- Fit the throttle twistgrip (7691 or 8899). On the later handlebar insert 28 *new* 1/8" balls (5608) into the greased ball race in the handlebar.

* Put the throttle twistgrip in place onto the right-hand handlebar.

* Fit the spring (7460) and for the later handlebar the steel and fibre washers (8789 and 8790) and the distance tube (8888).

* Fit the hooked lock pin (7211).

* Fit the throttle twistgrip stop screw (7488).

- For the early handlebar: fit the speedometer (7386 or 8366) instrument lighting (4496) if applicable, and ammeter (7665) or ignition warning light (8290).

- Fit the clutch and brake levers (7198) to the pivot bosses on the handlebar and secure them each with two screws (7204 or 7204-2).

Fitting the handlebar

_ Place the assembled handlebar on the forks and fork stem.

- On the early handlebar, fit the two stanchion nuts (7260) and washers (7527). There are two fibre washers (7752) which are to be fitted *under* the handlebar. On the later handlebar, fit the two stanchion bolts (9304) and washers (9303).

- Fit the stem nut washer (7576 or 9302) and the fork stem nut (7172) with the special tool (N 17/9004) or a 27mm spanner. (For frame numbers 1301 to 2600 use a 19 mm spanner).

- Fit the lighting switch (7345 or 8880) by means of the two screws (7834) if this has not yet been done. Make sure that both the lighting twistgrip and the switch rotor are in the "off" position. This means that the twistgrip has to be closed (turned fully forward) and the switch rotor has to be positioned so that no wiper contacts are in circuit.

- On the early handlebar fit the horn (7384) using two bolts (7265) and connect the leads. On the later handlebar, connect the leads from handlebar to the horn (7384).

- Fit the clutch cable (7672, 8413 or 8944) and brake cable (7671, 8248, 8707 or 8945) with their adjustment sleeve nuts (7208), rollers (7207), and if fitted, circlips (7198) to the levers (7198).

- Attach the throttle cable (8585 or 8927) to the hooked cross-pin (7211) in the throttle twistgrip.

- Fit the steering damper (7648, 7648-2 or 8897) and spring (3864). Put the bottom disc and nut (7647 or 9308), friction disc (7177) and the steering damper plate (7179-2 or 7179) in place under the fork yoke.

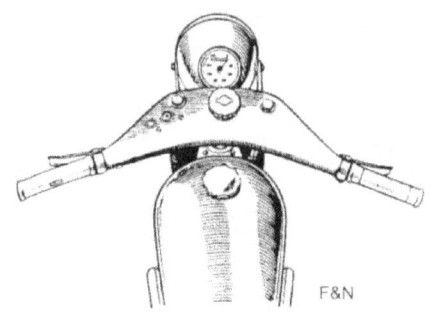

Repairing the handlebar

Accident damage to the handlebar usually takes the form of bends or fractures. Repairing the handlebar has to be done with care. If the handlebar is bent beyond factory specifications, it might be possible to correct this. (Check illustrations G7454 and G8875 in the digital drawings archive, Plan A).

A handlebar can remain safe for use after welding repairs but the option of replacement should be considered. On early handlebars welded reinforcement to the underside front is sometimes found. If this welding is correctly done the handlebar remains fit for use.
Slack in the clutch or brake levers (7198) is caused by wear in the shouldered pivot screws and/or of the holes in the levers. The holes can be drilled out oversize and

bushes then pressed into the levers. Or the pivot screws can be replaced by special collared versions.

If the lighting twistgrip (7692 or 8893) is not stable at a setting, remove the threaded plug (7650) securing the spring (7074) and ball (3845). Inspect the ball. If it is defective in any way it must be renewed. Press some grease into the hole, fit a new 5/16" ball, and refit the spring and plug. If the slack is caused by the ball having worn a groove between the detents of the twistgrip, then the twistgrip must be either replaced or repaired by welding and reprofiling the detents.

Saddle and pillion seat

There have been two different suspension systems for the saddles and pillion seats of Nimbus-C motorcycles, using either coil springs or rubber bands in tension.

Removing the coil-sprung saddle

- Unfasten both coil springs (7406 or 7407) at the frame.

- Remove the saddle pivot bolts.

- Free the saddle from the frame. Set aside the two pivot bushes (7596).

- Remove the coil springs from the saddle.

- Unclip the saddle cover (7669).

Detach the saddle springs (7595)

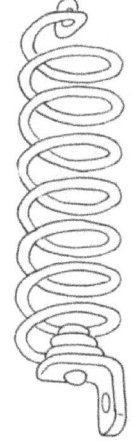

Fitting the coil-sprung saddle

Follow the removal instructions in reverse order. The saddle has 20 springs (7595). The two adjacent outer springs on each side share the same outside hole of nose bracket.

Removing
the rubber-suspended saddle

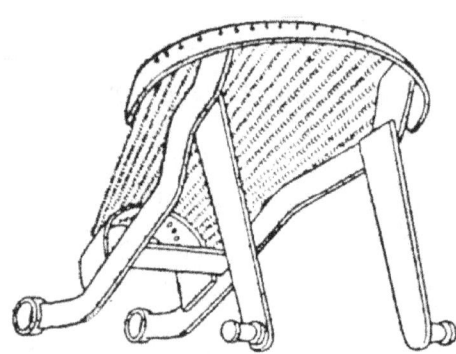

- Remove the nuts (7259) from the two saddle pivots (9851).

- Push the pivots in, out of the rubber bushes (9851), and remove the bushes from the pivot eyes of the saddle.

- Remove the two rubber suspension bands (9682 or 10103) and lift the saddle from the frame.

- Unclip the saddle cover (7669).

- Detach the saddle springs (7595).

- Remove the two suspension bobbins (9640) with buffers (9857).

Fitting
the rubber-suspended saddle

Follow the removal instructions in reverse order, except that the rubber suspension bands have to be fitted last. The saddle has 20 springs (7595). The two adjacent outer springs on each side share the same outside hole of the nose bracket. Apply some silicone lubricant to the saddle pivot rubber bushes.

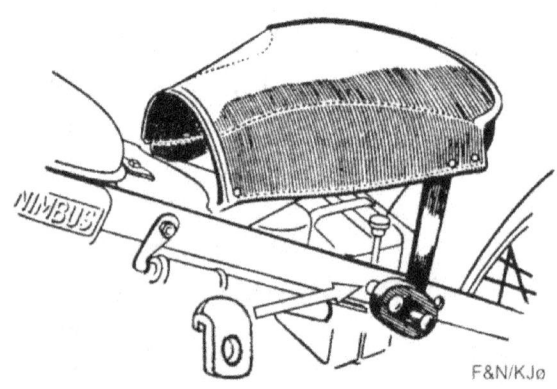

Removing
the coil-sprung pillion seat

- Unfasten both coil springs (7406 or 7407) at the mudguard stay.

- Remove the spindle (7816) of the combined rear mudguard and pillion seat pivot. Leave the

190

mudguard in place, held by the rear wheel spindle nuts.

- Lift the pillion seat from the motorcycle. Set aside the two pivot bushes (7817).

- Remove the coil springs (7406 or 7407) from the pillion seat.

- Unclip the pillion seat cover (7669).

- Detach the pillion seat springs (7595).
- Remove the pillion hand grip (7812-2).

Fitting
the coil-sprung pillion seat

Follow the removal instructions in reverse order. The pillion seat has 20 springs (7595). The two adjacent outer springs on each side share the same outside hole of the nose bracket. Make sure that the rear mudguard is correctly in place when fitting the pivot spindle (7816).

Removing
the rubber-suspended pillion seat

- Remove the through bolt (9646) of the combined rear mudguard and pillion seat pivot. Leave the mudguard in place, held by the rear wheel spindle nuts.

- Remove the two rubber suspension bands (9682 or 10103) and lift away the pillion seat.

- Remove the rubber bushes (9856) from the pivot eyes of the pillion seat.

- Unclip the pillion seat cover (7669).

- Detach the pillion seat springs (7595).

- Remove the pillion hand grip (7812-2).

- Remove the two suspension bobbin brackets (9680 and 9681). Identify the right and left brackets before removal as they are not interchangeable.

**Fitting
the rubber-suspended pillion seat**

Follow the removal instructions in reverse order. The pillion seat has 20 springs (7595). The two adjacent outer springs on each side share the same outside hole of the nose bracket. Make sure that the rear mudguard is correctly in place when fitting the pivot spindle (7816). Apply some silicone lubricant to the saddle pivot rubber bushes.

**Fitting
rubber suspension bands to the pillion seat**

Suspension bands are marked 80kg or 100kg. Fitting can be done as follows:

- Place the suspension bands on each side fitting around the bobbin of the pillion seat and around both bobbins of the mudguard stay brackets, that is, <u>around all three bobbins</u> on each side.

- Stretch the bands by having someone sit on the pillion seat.

- Work the *upper* run of each band *under* the middle bobbin (the lower bracket bobbin) using a lever or a strap to assist.

- Apply silicone lubricant to bobbins and bands.

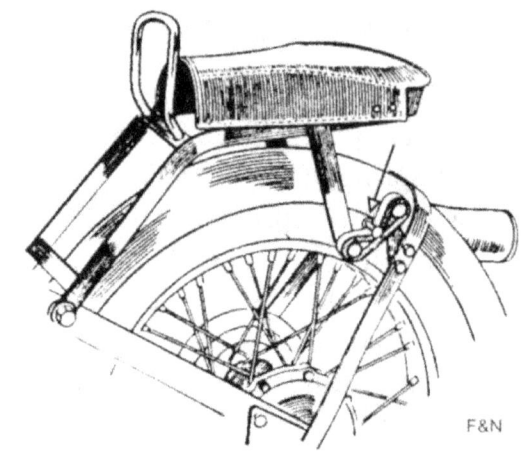

Repairs

Replace springs and rubber suspension bands as necessary - preferably *before* they break.

The fitting holes, especially the rear fitting holes, for the 20 springs (7595) on both saddle and pillion seat will wear. If any holes have worn right through they can be repaired by welding.

Slack in, or noise from a coil-sprung saddle or pillion seat can be a result of wear in pivot bushes (7596 or 7817), pivot spindle or pivot through-bolt, or saddle or seat mounting arms. Holes in the mounting arms can be welded up and re-drilled to the correct diameter or can be drilled oversize and used together with correspondingly oversized bushes.

Gear changing

Gear changing on the Nimbus-C is by means of either a hand-change gear lever or a foot-change gear pedal. Up to production number 7500, either can have been fitted. From number 7501 on, all motorcycles were fitted with the foot-change mechanism.

Removing the gear lever (hand-change equipped motorcycles)

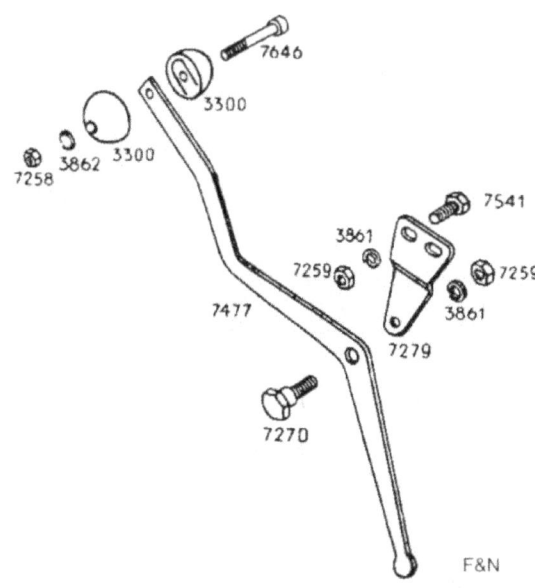

- Remove the nut (7258) and spring washer (3862) from the gear lever pivot (7270) and remove the pivot bolt from the gear lever (7477) and the gear lever pivot bracket (7279).

- Remove the gear lever (7477) with knob (3300) from the gearbox selector shaft (7819) and withdraw it through the gear lever gate between the tank and saddle.

- Remove the gear lever gate (7279) which is secured by four screws (7541), spring washers (3861), and nuts (7259).

- Remove the gear lever knob securing bolt (7646), spring washer (3862), and nut (7258) and set aside the two parts of the knob.

Fitting the gear lever (hand-change equipped motorcycles)
Follow the removal instructions in reverse order. See also "Adjusting the gear lever" below.

Removing the gear pedal assembly (foot-change equipped motorcycles)

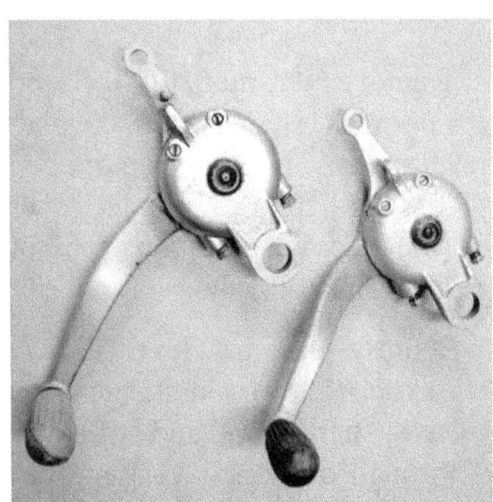

- Remove the nut from the shouldered bolt (8405) of the gearbox selector shaft (8408 or 9221) and take out the bolt.

- Remove the foot rest nut (7261) that secures both the footrest (7357) and the foot-change mechanism and remove the footrest. This frees the foot-change mechanism.

Dismantling the foot-change mechanism

- Remove the two adjusting screws (7191) and springs (8523) from beneath the foot change mechanism (8395).

- Hold the head of the pivot bolt (8404) in a vice and remove the nut (8522) at the back of the assembly.

- Remove the pivot bolt (8404) and the gear pedal (8418), which is integral with the back plate of the foot-change housing. Take the pawl disc (8397 or 9556) from the housing (8395).

- Prise the pedal return spring (8402) and the spring abutments (8407) out of the housing. Take care, as the spring is compressed and can flick out the abutments with some force and for some distance.

- Remove the pawl stop plate (8401) which is kept in place by two screws (5253).

- Leave the two pawls and pivots (8411 and 8412) in place unless they have to be renewed. If this is necessary, remove the pawl spring (8403) from between the pivots. Use a punch to drive the pivots (8406) out of the back plate. The pivots (8406) are brass and can normally be reused if handled with care.
- Remove the pedal rubber (8409).

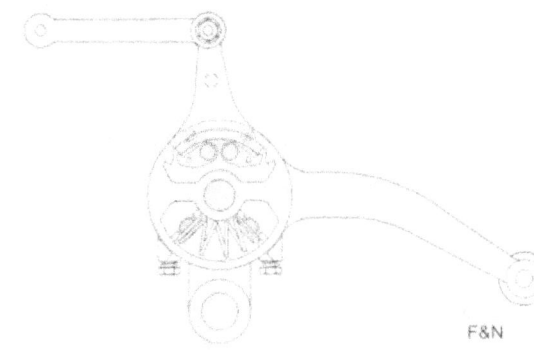

F&N

Assembling the foot-change mechanism

- If the two pawls and pivots (8411 and 8412) have been removed, fit them and put back the pawl spring (8403). Grease all parts.

- Attach the pawl stop plate (8401) to the housing by the two screws (5253).

- Fit the two adjustment screws (7191) and springs (8523).

- Fit the pedal return spring (8402) and the two spring abutments (8407) between the abutments in the housing.

- Pack the housing with grease.
- Assemble the housing, pawl disc, and back plate with integrated gear pedal, then tighten the central pivot bolt (8404) and nut (8522). *Make sure* the pawl disc is correctly oriented, so that the gear indicator pin is to the outside.

Repairing the foot-change mechanism

The foot change mechanism is prone to wear and one or more of the following repairs may be required:

Renewing the central pivot bolt (8404):
- When the central pivot bolt needs to be renewed, the hole in the foot change mechanism housing is likely to be worn out as well. In that case it pays to drill out the housing and fit a bush. A Nimbus small end bush (10188) can be useful for this purpose. It has the needed inner diameter of 16mm.

Replacing the pawls (8411 and 8412):
- The two pivots (8406) must be driven out and new right and left pawls (8411 and 8412) fitted to the pivots. Secure the pivots with Loctite if necessary.

Replacing of the pedal return spring and the spring abutments:
- The pedal return spring (8402) and the two spring abutments (8407) tend to wear more on one side. They can be turned over and re-used, but it is better to renew the parts.

Replacing the pawl stop plate (8401):

- If wear marks are clearly visible, the pawl stop plate has to be renewed.

Replacement of the peg (8410) for the gear pedal rubber:
- The peg (8410) for the gear pedal rubber is pressed into the pedal and riveted. If a new peg has to be fitted, the old one must be driven out.

Adjusting the hand-change mechanism

- Release the bolts (7541) that hold the gear change gate (7279) in place. The fixing holes are oval to allow the plate to be adjusted in the frame.
- Put the gearbox selector shaft (7819) in the neutral position, between 1st and 2nd gear.
- Place the gear lever (7477), fitted on the pivot bolt (7270), so that the rounded operating end is central in the slot of the selector shaft. Move the gate so that the lever is aligned with the neutral indication "0".
- Tighten the gear change gate and check that the change gate indication is correct in all gear positions.

Adjusting the foot-change mechanism

- Loosen the nut (7261) of the foot rest (7357).

- Put the selector shaft (7819 or 9210) in the neutral position between 1st and 2nd gear.

- Position the gear pedal (8418) so that the gear indicator pin (9556) aligns as closely as possible to the clutch cable abutment on the foot-change housing (8395). If the pawl disc (8397) does not have a gear indicator pin, align the index mark on the edge of the pawl disc as closely as possible to the visible forward edge of the pawl stop plate (8401).

- Fine adjustment of the gear indicator is by means of the adjusting bolts (7191) which bear on the frame. These allow the housing of the foot-change mechanism (8395) to be set so that in neutral, there is exact alignment of the gear indicator pin and the clutch cable abutment, or of the pawl disc index mark and the front edge of the pawl stop plate.

- Tighten the footrest nut.

Speedometer

Three different speedometers were used on the Nimbus-C: 'VDO' 60mm diameter (8294 or 8366) fitted in the handlebar, 'Smiths' 80mm diameter (9328), fitted by a bracket to the right headlamp support or more rarely at the centre of the handlebars, and finally 'VDO' 80mm diameter (9902) fitted in the headlamp.

Removing the speedometer

- Remove the speedometer cable (7753, 9329 or 9903). The outer speedometer cable has a captive knurled nut at each end allowing connection to the speedometer and the cable adapter of the front wheel speedometer drive housing.

- Detach the lead to the speedometer light. (For a headlamp-mounted see 3c below).

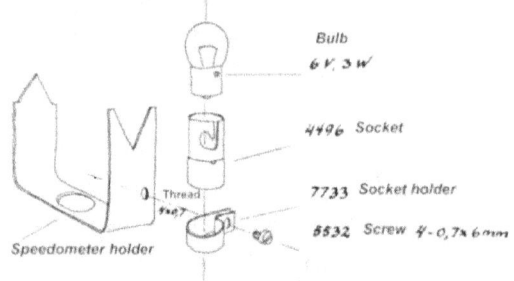

INSTRUMENT LIGHTING 1301 -2550

Remove the speedometer:

* For a VDO speedometer fitted in the handlebar, release the single nut securing the brace which retains the instrument and lift the speedometer (7386, 8294 or 8366) from the handlebar.
* Remove the 'Smiths' speedometer (9328) by removing the two nuts securing the retaining brace.
* Headlamp-mounted VDO speedometers require that the headlamp rim (9947), lens (9948), and reflector (9949) with bulb holder be removed. Then detach the lead (#6/yellow) to the speedometer light, release the two finger nuts of

the retaining brace and remove the speedometer.

Fitting the speedometer

Follow the removal instructions in reverse order. Be careful to install the inner cable correctly in the outer, so that the locating collar at one squared section of the cable is at the speedometer end.

Dismantling the speedometer

Owner-dismantling of speedometers is not recommended. Speedometers in which the glass is sealed by a crimped-on rim require for dismantling that the rim be prised off, after which it cannot normally be re-used. The Smiths instrument has the glass retained by a threaded chrome rim which can be unscrewed and refitted.

Maintenance and repair of the speedometer

Maintaining speedometers is primarily about keeping the speedometer cable and drive gears well-greased. The cable should be disconnected from the speedometer, the inner cable removed, and grease applied to its entire length. The cable adapter and speedometer pinion should be removed from the speedometer drive housing at the front wheel and the housing packed with grease.

A broken speedometer inner cable is easily replaced. Make certain that the replacement cable is of the correct length and install it from the speedometer end of the outer cable so that the locating collar of the inner cable bears against the speedometer.

Overhaul of speedometers is a job for specialists, whether it concerns renewing the dial, repairing or resetting the odometer mechanism, or calibrating the speed reading.

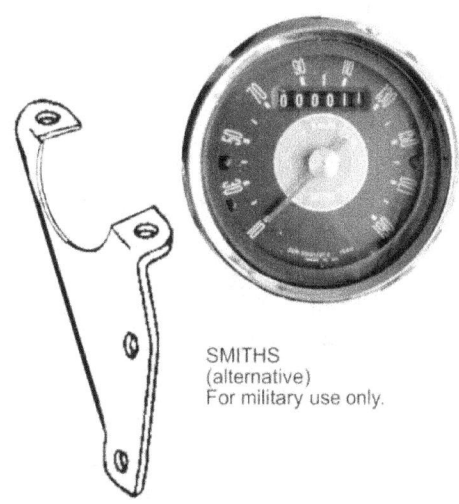

SMITHS
(alternative)
For military use only.

Valve enclosure

In 1956, new Nimbus-C motorcycles were fitted with eight individual valve enclosures which could be retro-fitted to older machines. These enclosures proved to be technically problematic despite the many modifications which were made, and few machines so equipped are now in use.

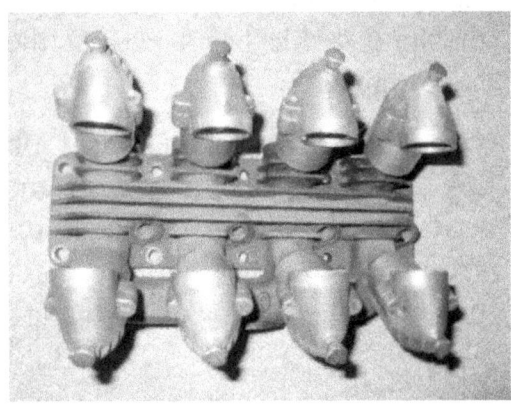

Sidecars

The following sidecar types were available for fitting to Nimbus-C motorcycles:

Tube frame without brakes:
* Supplied in 1934 and 1935,

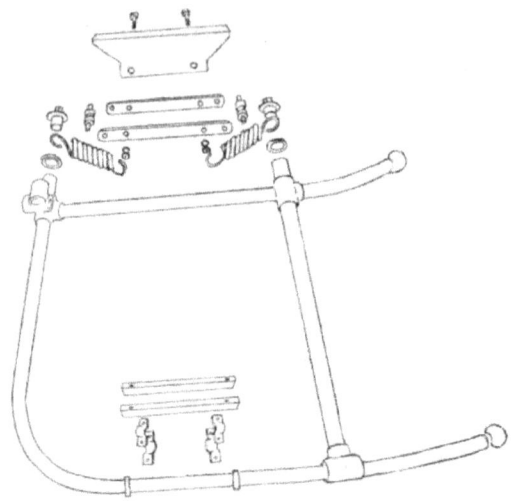

Flat steel frame without brakes:
* Type U (1935 to 1938)

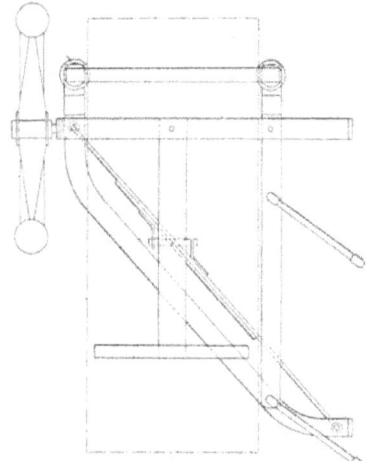

Tube frame with brakes:
* Type R (1938 and 1939)

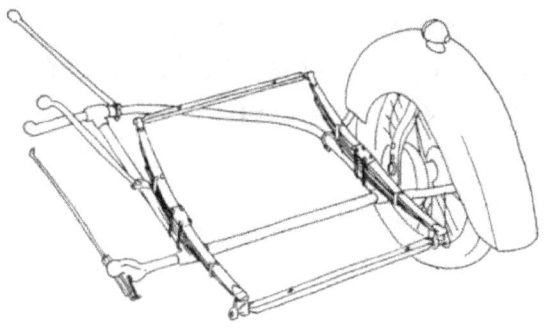

* Type R/RB (1940 to 1947)
* Type RB (1948 to 1966).

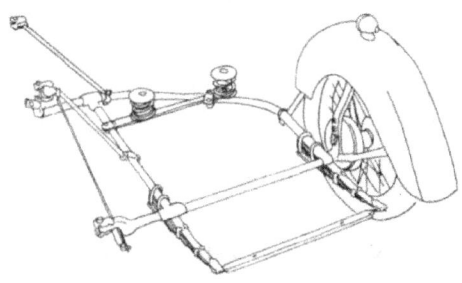

Sidecar maintenance and repair

For sidecars fitted with a braked wheel, regularly grease the brake cam pivot (8497) by means of the grease nipple (7649) fitted to the sidecar frame close to the lower rear ball coupling.

- Remove and grease the wheel bearings annually, or after 5,000km.

- Spray the leaf spring regularly with protective oil to prevent rusting. (Spray-on motorcycle chain oil is highly suitable for this purpose, being designed to penetrate and then to dry).

- Regularly check the tightness of the four sidecar ball couplings (7969). Loose couplings are unsafe and are prone to rapid wear. Avoid over-tightening, which can result in stretched bolts or stripped threads.

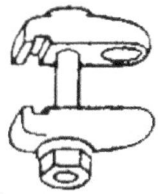

- Keep both the motorcycle and the sidecar brakes correctly adjusted (see below). For repair of brakes, see the section on brakes.

- The sidecar wheel spokes are laced as shown in the diagram - see the section on wheels.

- Wiring for the sidecar light must be in good condition and securely installed. The sidecar light normally relies on a good earth connection being made to the motorcycle frame through the ball couplings. It may be worthwhile installing an earth lead from the sidecar frame direct to an earth point on the motorcycle, particularly on a newly repainted sidecar outfit.

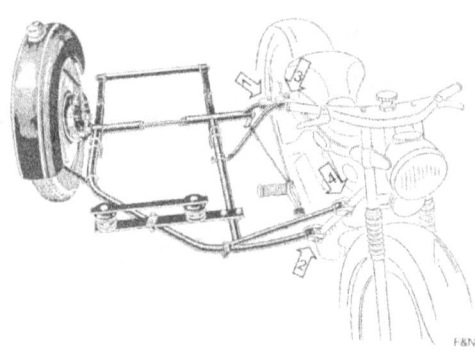

Adjusting the sidecar brake

- First, adjust the motorcycle rear wheel brake (see above).

- Check the effectiveness of the sidecar brake. Ride the motor cycle, with sidecar, at moderate speed on a straight, dry, and preferably traffic-free road and apply the brake firmly. The outfit should not pull to either side. If it is pulled to the right, the sidecar brake adjuster (9598) on the sidecar brake pull rod (8499) must be slackened. If it is pulled to the left, the sidecar brake adjuster must be tightened.

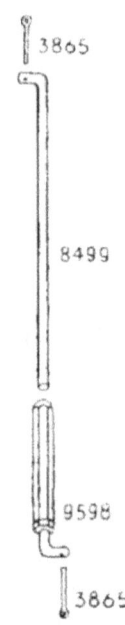

- Check the angle between the sidecar brake arm (8476) and the sidecar brake pull rod (8499). If this angle is more than 90° with the brake applied, remove the brake arm and reverse it so that the "S" stamping is visible. If this has already been done, the brake should be dismantled for lining replacement and possibly drum re-machining.

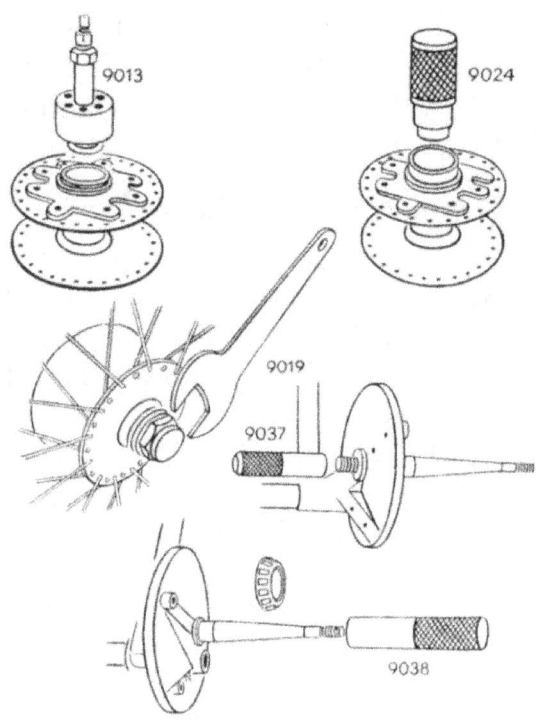

Adjustments

Adjusting the rear wheel bearings
See Rear wheel

Adjusting the brakes
See Brakes

Adjusting the crownwheel and pinion (final drive gears)
See Rear wheel

Adjusting the carburettor
See Carburettor

Adjusting the gearchange mechanism
See Gear changing

Adjusting valve clearances

Adjusting valve clearances is always done with a *cold* engine.

More possibilities available:
One way, where the engine must be turned <u>eight</u> times and another, where the engine must be turned only <u>twice.</u>

The ´Eight-position method´:

- Clean the rockers, adjustment screws, and valve spring collars.

- Remove the spark plugs to facilitate turning the engine to the correct position for valve adjustment.

- Turn the engine with the kick starter and note when a valve is fully open (that is, when the rocker has fully depressed the valve). For either the intake or the exhaust side of the engine, this procedure ('the rule of 5') can be followed:

* When valve #1 (front cylinder) is fully open, valve #4 (rear cylinder) can be adjusted.
* When valve #4 is fully open, valve #1 can be adjusted.
* When valve #2 is fully open, valve #3 can be adjusted.
* When valve #3 is fully open, valve #2 can be adjusted.

Adjusting valve clearances <u>with</u> feeler gauges

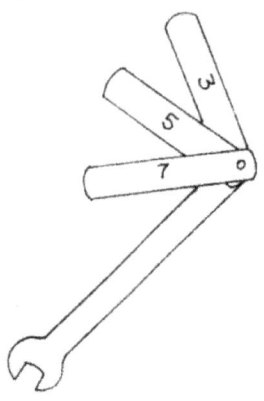

The clearance (between valve stem and the rocker adjuster with a *cold* engine) must be 0.3mm for all inlet valves and 0.7mm for all exhaust valves.
- Loosen the locknut of the rocker adjuster with a 10mm spanner, insert a feeler gauge between the adjuster and the valve stem. Set the rocker adjuster so that the feeler gauge can just slide between it and the valve stem with some friction being felt.

- Hold the rocker adjuster with a screwdriver to prevent it turning and tighten the locknut. *Do not over-tighten.* Check the valve clearance again to confirm that it has not changed while tightening the locknut.

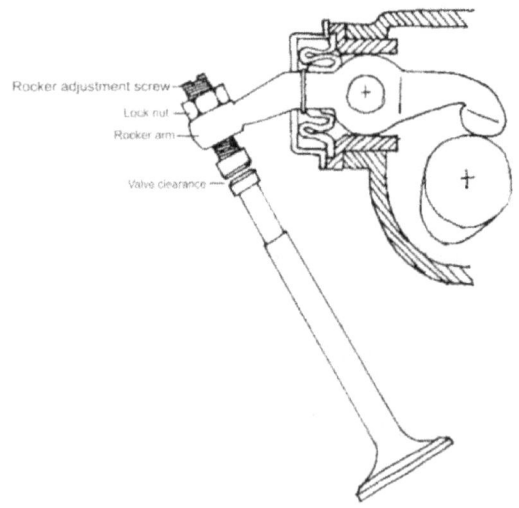

Adjusting valve clearances <u>without</u> feeler gauges

The clearance (between valve stem and the rocker adjuster with a *cold* engine) must be 0.3mm for all inlet valves and 0.7mm for all exhaust valves.

- Loosen the locknut of the rocker adjuster with a 10mm spanner.

- Screw down the rocker adjuster until it contacts the valve stem (zero clearance).

- For the <u>inlet</u> valves, back off the rocker adjuster exactly <u>half a turn</u>. Because of the thread pitch this will give a clearance of 0.375 mm. Hold the rocker adjuster with a screwdriver to prevent it turning and tighten the locknut. *Do not* over-tighten.

- For the <u>exhaust</u> valves, back off the rocker adjuster exactly one turn. Because of the thread pitch this will give a clearance of 0.75mm. Hold the rocker adjuster with a screwdriver to prevent it turning and tighten the locknut. *Do not* over-tighten.

The ´Two-position method´:

Remove the spark plug from cylinder number 1.

- Place a 25mm board under the centre stand and pull the motorcycle onto its stand. Turning the rear wheel by hand, engage 3rd gear.

- Turn the engine over by means of the rear wheel. When the intake valve of cylinder number 1 opens and then closes, the piston is approaching TDC (top dead centre) on the compression stroke.

- Use a 'feeler' through the spark plug hole (a pencil or an old bicycle spoke will do) to sense when TDC on cylinder number 1 is reached exactly. At this point four of the eight valves are fully closed, and these clearances can be set:

* Cylinder 1: Both intake and exhaust
* Cylinder 2: Exhaust only
* Cylinder 3: Intake only

Using the same technique, turn the engine one complete revolution (360°) until TDC in cylinder 1 is reached again. Now the other four valves will be fully closed, ready for the remaining clearances to be set:
* Cylinder 2: Intake only
* Cylinder 3: Exhaust only
* Cylinder 4: Both intake and exhaust.

Or ...

- Remove the spark plug from cylinder number 1.

- Place a 25mm board under the centre stand and pull the motorcycle onto its stand. Turning the rear wheel by hand, engage 3rd gear.

- Turn the engine over by means of the rear wheel. When the intake valve of cylinder number 1 opens and then closes, the piston is approaching TDC (top dead centre) on the compression stroke, when both valves will be fully closed. Use a 'feeler' through the spark plug hole (a pencil or an old bicycle spoke will do) to sense when this position is reached exactly. At this point the four valve clearances ticked in the table below can be set:

Valves to be adjusted at Cylinder 1 TDC compression stroke

	Cylinder 1	Cylinder 2	Cylinder 3	Cylinder 4
Exhaust	✓	✓	✗	✗
Intake	✓	✗	✓	✗

- Using the same technique, turn the engine one complete revolution (360°) until TDC in cylinder 1 is reached again. Now the other four valves will be fully closed, ready for clearances to be set: These are shown crossed in the table above.

(The table appears in Honda service publications for the 350-900 4-cylinder range)

Adjustment of the contact breaker

- Turn the engine so that the contact breaker points are fully open, and the fibre heel of the opening contact is on one of the four lobes of the contact breaker cam (7400).

- Slightly loosen the upper securing screw (7399) of the fixed contact in the distributor and turn the lower eccentric screw (7398) to set the contact breaker gap to 0.7mm, measured by feeler gauge.

- Secure the fixed contact by tightening the upper screw (7399).

- Turn the engine with the kickstarter and check the contact gap for all four lobes of the cam. Make sure the distributor is pressed fully home in the camshaft housing while making these checks.

Adjusting the ignition timing

- Loosen the securing screw (5350) in the slot of the timing indicator plate attached to the distributor and turn the distributor to align the ' S ' mark with the screw. *The ' S ' indicates <u>advanced</u>. (Danish: 'sent' = late).*

- Take off flywheel access cap and turn the engine until the ' I ' (ignition) mark on the flywheel is seen to be central in the hole. This is the ignition point for either cylinder 1 or 4. *Note that this is NOT the top dead centre (TDC) mark.*

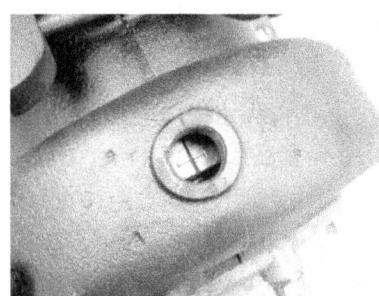

- Remove the distributor lead and connect a test lamp between lead and the distributor terminal.

-There are many different opinions about what the best ignition setting is. An experienced mechanic can set the ignition by ear. There were no written instructions from the factory to their authorised workshops on how the ignition ought be set.

Either: Rotate the flywheel until the ignition mark ' I ' is aligned with the right edge of the access hole in the flywheel housing.

Or: Rotate the flywheel, until the ignition mark ' I ' has moved 22 to 27 mm past the hole. This will result in a later (retarded) and perhaps better setting.

- Switch on the ignition, turning on both the charge warning light on the handlebar and the test lamp

- Turn the distributor back (against normal engine rotation) in the direction of the ' T ' mark until the test lamp goes out, indicating that the contact breaker points have just opened. Normally, this will happen when the screw is halfway the ' S ' and the ' T ' mark.

- Put a spring washer under the securing screw (5350) and tighten it.

Adjusting clearances in the valve timing gear

The valve timing gear consists of four bevel gears: at the crankshaft (7053), the lower and end of the dynamo armature (7147), the upper end of the dynamo (8056 or 8056-2), and at the camshaft (7094 or 7094-2).

The drawings archive shows tooling for the adjustment (9043 and 9044, 'Adjustment gauge for bevel gear wheels'). However it is possible without this tooling to correctly adjust the two clearances: between the crankshaft and dynamo bevel gears, and between the dynamo and camshaft gears.

Adjusting the clearance between the crankshaft and dynamo bevel gears

Before fitting the crankshaft, it is important to know that the clearance between the crank shaft and dynamo bevel gears, when assembled, will be correct.

- Fit the assembled dynamo to the cylinder block, without the use of any liquid gasket.

- Lay the crank shaft, fitted with its main bearings and bevel gear, in the cylinder block.

- Tighten the main bearing caps without the use of locking washers.

- Turn the crank shaft and note the backlash between the two gears. This should be about 0.2mm. Place shims between the crankshaft gear and the main bearing until there is zero clearance and then remove 0.15mm to 0.20mm of shims.

Adjusting the clearance between the dynamo and camshaft bevel gears

- Fit the assembled dynamo to the cylinder block, without the use of any liquid gasket.

- Fit the cam shaft housing with the camshaft in place but not necessarily with rockers fitted. Tighten the four fixing bolts.

- Place shims under the dynamo upper gear until there is zero clearance and then remove 0.15mm to 0.20mm of shims. Note that if there is no clearance *without* any shims, 0.2mm must be machined off the face of the front camshaft bush. Do not attempt to resolve this issue by shimming up the camshaft housing. That will result in excessive valve guide wear.

Helical cut bevel gears
The procedure for adjusting clearances with helical cut bevel gears is essentially the same.

Tyre pressures:

Front wheel	Solo motorcycle	1.6 bar /23psi
	Motorcycle with sidecar	1.8 bar /26psi
Rear wheel	Solo motorcycle	1.8 bar /26psi
	Motorcycle with sidecar	1.9 bar /28psi
Sidecar wheel		1.6 bar /23psi

Torque settings:

	ft/lbs	Nm
Main bearingcap nuts	45	60
Connecting rod big end nuts	25	34
Cylinder head fixing bolts-/nuts	25	34
Crown wheel to hub bolts	25	34
Brake drum to hub bolts	25	34
Fork yoke clamp nuts	25	34
Spark plugs	15	20
Flywheel to crankshaft nut	Tighten hard	Tighten hard
Dynamo armature nuts	Tighten hard	Tighten hard

Lubricating Nimbus motorcycle

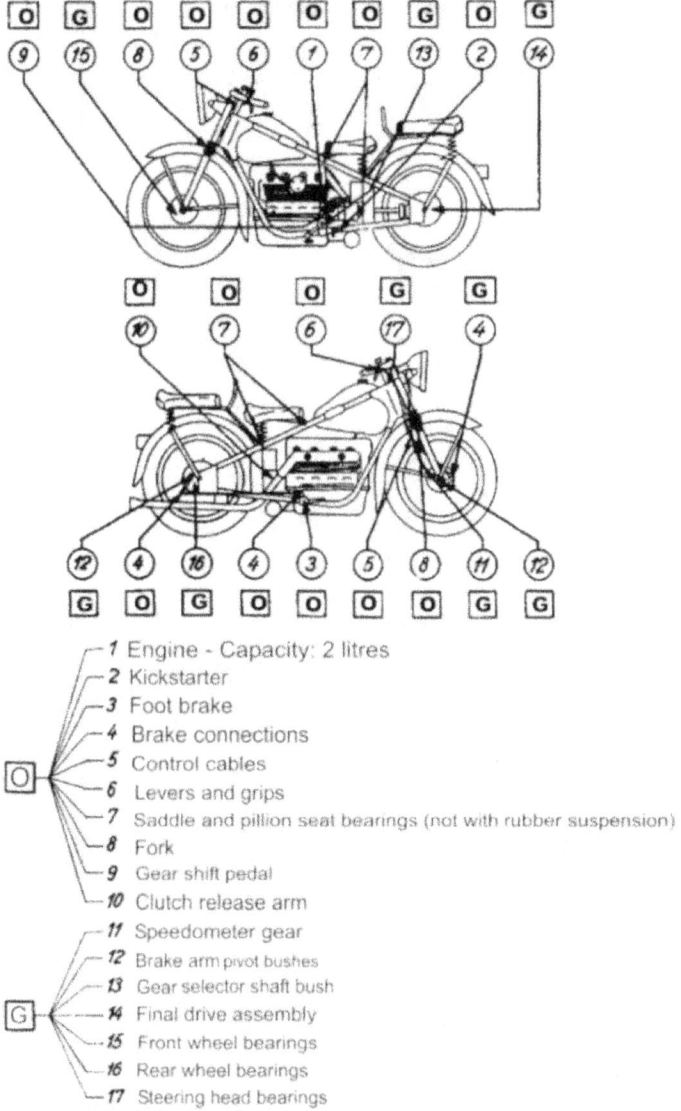

1. Engine - Capacity: 2 litres
2. Kickstarter
3. Foot brake
4. Brake connections
5. Control cables
6. Levers and grips
7. Saddle and pillion seat bearings (not with rubber suspension)
8. Fork
9. Gear shift pedal
10. Clutch release arm
11. Speedometer gear
12. Brake arm pivot bushes
13. Gear selector shaft bush
14. Final drive assembly
15. Front wheel bearings
16. Rear wheel bearings
17. Steering head bearings

O Oil, winter SAE 20
O Oil, summer SAE 30 - 40   G Grease

Efter: HÅNDBOG for hærens motorførere
(Nimbus motorcykel) 1951

Nimbus Paint Colours - and Modern Equivalents

Colour	Colour Name Original name	Colour Name RAL system	RAL-Number	Year Used
Black 1	Black	Jet Black	RAL-9005	1934-60
Red 6	Red	Wine Red	RAL-3005	1935-45
Red 7	Bordeaux	Purple Red	RAL-3004	1952-59
Green 2	Green	Pine Green	RAL-6028	1935-45
Green 3	Withered Green	Reseda Green	RAL-6011	1954-60
Green 4	Deep Sea Green	Blue Green	RAL-6004	1954-60
Blue 13	Tivoli Blue	Steel Blue	RAL-5011	1937-45
Blue 14	Blue	Green Blue	RAL-5001	1946-50
Yellow 8	Ivory/Yellow	Ivory	RAL-1014	1939-45
Grey 10	Lavender/Grey	Grey Beige	RAL-1019	1939-43
Grey 11	Polychromatic Grey	Grey Aluminum	RAL-9007	1943-45
Other	**Defence Colours**			
Green 5a	Olive/Grey	Olive Grey	RAL-7002	1934-45
Green 5b	Grey/Olive	Grey Olive (matt)	RAL-6006	1945-60
Grey 12	Civil defence Grey	Traffic Grey A	RAL-7042	1950-60
Yellow 9	Danish Post Yellow replaced by	Maize Yellow	RAL-1006	1934-60
Electrostatic Enamel Bright Silver (used on the lower part of the fuel tank and other misc parts)		White Aluminium	RAL-9006	1934-60

www.ingramcontent.com/pod-product-compliance
Lightning Source LLC
Chambersburg PA
CBHW082326220526

45470CB00008B/2414

Knud Jørgensen
NIMBUS - Maintenance

English Edition 2018
Translation: Ben Geutskens and Charles Duffill

This book is intended as a guide to the maintenance and repair of the Danish Nimbus type C motorcycles built between 1935 and 1959.
It is by large limited to those operations which a skilled owner can do or can have done.

The many changes to Nimbus-C between 1934 and 1959 are described in:
Knud Jørgensen: NIMBUS - Technical Development
by Books on Demand GmbH, Copenhagen, Denmark 2016.

www.bod.dk